Springer Proceedings in Physics

Volume 282

Indexed by Scopus

The series Springer Proceedings in Physics, founded in 1984, is devoted to timely reports of state-of-the-art developments in physics and related sciences. Typically based on material presented at conferences, workshops and similar scientific meetings, volumes published in this series will constitute a comprehensive up-to-date source of reference on a field or subfield of relevance in contemporary physics. Proposals must include the following:

- Name, place and date of the scientific meeting
- A link to the committees (local organization, international advisors etc.)
- Scientific description of the meeting
- List of invited/plenary speakers
- An estimate of the planned proceedings book parameters (number of pages/articles, requested number of bulk copies, submission deadline).

Please contact:

For Americas and Europe: Dr. Zachary Evenson; zachary.evenson@springer.com
For Asia, Australia and New Zealand: Dr. Loyola DSilva; loyola.dsilva@springer.com

Rajendra Nath Patra
Editor

Advanced Radiation Detector and Instrumentation in Nuclear and Particle Physics

Proceedings of RAPID 2021

Springer

Editor
Rajendra Nath Patra
University of Jammu
Jammu, India

ISSN 0930-8989 ISSN 1867-4941 (electronic)
Springer Proceedings in Physics
ISBN 978-3-031-19267-8 ISBN 978-3-031-19268-5 (eBook)
https://doi.org/10.1007/978-3-031-19268-5

This Springer imprint is published by the registered company Springer Nature Switzerland AG
The registered company address is: Gewerbestrasse 11, 6330 Cham, Switzerland

Organization

Scientific Advisory Committee

Bhasin, Anju, University of Jammu
Bhatnagar, Vipin, Panjab University
Bheesette, Satyanarayana, TIFR
Biswas, Saikat, Bose Institute
Chattopadhyay, Subhasis, VECC
Deshpande, Abhay, Stony Brook University
Ghosh, Tilak, VECC
Graafsma, Heinz, DESY
Mohanty, Bedangadas, NISER
Mukhopadhyay, Supratik, SINP
Musa, Luciano, CERN
Nayak, Tapan, NISER, CERN
Sharma, Archana, CERN
Singh, Bhartendu, Banaras Hindu University

Organizing Committee Members

Arya, Sandeep, University of Jammu
Bala, Renu, University of Jammu
Bamzai, K. K., University of Jammu
Bharti, Arun, University of Jammu
Bhasin, Anju, University of Jammu
Devi, Rani, University of Jammu
Gupta, Anik, University of Jammu
Gupta, Ramni, University of Jammu
Gupta, Vivek, University of Jammu

Kant, Rajni, University of Jammu
Mahajan, Sanjay, University of Jammu
Nayak, Saroj Kanta, University of Jammu
Padha, Naresh, University of Jammu
Patra, Rajendra Nath, University of Jammu
Sambyal, Sanjeev Singh, University of Jammu

Program Committee

Bhasin, Anju, University of Jammu
Biswas, Saikat, Bose Institute
Patra, Rajendra Nath, University of Jammu
Sambyal, Sanjeev Singh, University of Jammu

Preface

The international workshop on Advanced Radiation Detector and Instrumentation in Nuclear and Particle Physics (Proceedings of RAPID 2021) was organized by the University of Jammu on 25–29 October 2021. The workshop was held online due to the COVID-19 restrictions. The focus of this workshop was on the contemporary and technologically challenging topics of micro-pattern gas detectors, resistive plate chambers, silicon detectors and their application in high energy physics experiments, detector technologies in nuclear physics, detector simulation, societal applications, and future prospects.

The Proceedings of RAPID 2021 aimed to bring researchers, scientists, professors, and students working in the field of designing and testing detectors for nuclear and particle physics experiments and applications to present new results and technologies, discuss and exchange ideas. The workshop is the first of its kind in India to bring researchers working in different parts of the world on detector technology for high-end talks, fruitful discussions, and debates. The workshop attracted more than 150 participants including a large number of Ph.D. students and young scientists who made the workshop quite vibrant and lively.

Professor Manoj Dhar, Hon'ble Vice Chancellor, University of Jammu, inaugurated the workshop with the keynote address in the presence of invited guests and participants. The scientific deliberations included invited lectures, contributed presentations, talks by young researchers as well as laboratory tours to highlight the technologies from MPGD, ALICE, INO, CBM, Dark Matter, Calorimeter, and Particle Therapy. The workshop ended with the announcement of the three best-contributed talks and concluding remarks by the Chairman of the workshop, Prof. Sanjeev Singh Sambyal. The workshop concluded with an evening lecture on "History of Detector Development and Future Perspective in India" by Prof. Naba K. Mondal.

The scientific and organizational success of Proceedings of RAPID 2021 was the result of work of the many distinguished colleagues. Members of the Scientific Advisory Committee played a vital role in the selection of the invited speakers. The contributions of the Organizing Committee in different aspects are much appreciated. The papers for the proceedings are peer-reviewed with the help of many volunteer reviewers in the field, to which the organizer wishes to extend their gratitude. Finally, the Speakers, the Session Chairs, and the Participants, all together made the workshop quite special.

More information about the workshop, lectures, and talks can be found at https://indico.cern.ch/event/1068553/.

Jammu, India

Rajendra Nath Patra
Guest Editor

Contents

Contributors

Rama Prasad Adak Department of Physics, Taki Government College, Taki, West Bengal, India

Shuaib Ahmad Khan VECC, HBNI, Kolkata, India

Suraj Ali Saha Institute of Nuclear Physics, Kolkata, India;
Department of Physics, Jadavpur University, Kolkata, India;
Muralidhar Girls' College, Kolkata, India

Jyoti Babbar Department of Physics, Panjab University, Chandigarh, India

Yong Ban School of Physics, Peking University, Beijing, China

Shiba Prasad Behera Bhabha Atomic Research Centre (BARC), NUCLEAR PHYSICS DIVISION (NPD), Trombay, Mumbai, MH, India

Vipin Bhatnagar Department of Physics, Panjab University, Chandigarh, India

Rupamoy Bhattacharyya Institute of Physics, Bhubaneswar, Odisha, India

Mouli Chaudhuri School of Physical Sciences, National Institute of Science Education and Research, Jatni, India;
Homi Bhabha National Institute, Training School Complex, Mumbai, India

Sushil S. Chauhan Department of Physics, Panjab University, Chandigarh, India

Mala Das Saha Institute of Nuclear Physics, Kolkata, India

S. Das School of Physical Sciences, National Institute of Science Education and Research, Jatni, India;
Homi Bhabha National Institute, Training School Complex, Mumbai, India

Subhendu Das Saha Institute of Nuclear Physics, A CI of Homi Bhabha National Institute, Kolkata, West Bengal, India

Jaydeep Datta Saha Institute of Nuclear Physics, A CI of Homi Bhabha National Institute, Kolkata, West Bengal, India

Roni Dey Bhabha Atomic Research Centre (BARC), NUCLEAR PHYSICS DIVISION (NPD), Trombay, Mumbai, MH, India

Nitish Dhingra Department of Soil Science, Punjab Agricultural University, Ludhiana, India

Yuvaraj Elangovan Department of High Energy Physics, Tata Institute of Fundamental Research, Colaba, Mumbai, MH, India

Riya Gaba Department of Physics, Panjab University, Chandigarh, India

Vishwajeet Jha Bhabha Atomic Research Centre (BARC), NUCLEAR PHYSICS DIVISION (NPD), Trombay, Mumbai, MH, India

Aera Jung School of Physics, Peking University, Beijing, China;
University of Hawaii, Honolulu, USA

R. Kanishka Applied Nuclear Physics Division, Saha Institute of Nuclear Physics, Kolkata, India;
Homi Bhabha National Institute, Mumbai, India

V. K. S. Kashyap School of Physical Sciences, National Institute of Science Education and Research, Jatni, India;
Homi Bhabha National Institute, Training School Complex, Mumbai, India

Amandeep Kaur Department of Physics, Panjab University, Chandigarh, India

Harjot Kaur Department of Physics, Panjab University, Chandigarh, India

Aafke Kraan INFN Pisa, Largo Bruno Pontecorvo, 3/Edificio C, Pisa, Italy

Shyam Kumar University and INFN, Bari, Italy

Nayana Majumdar Applied Nuclear Physics Division, Saha Institute of Nuclear Physics, Kolkata, India;
Homi Bhabha National Institute, Mumbai, India;
Saha Institute of Nuclear Physics, A CI of Homi Bhabha National Institute, Kolkata, West Bengal, India

G. Majumder Department of High Energy Physics, Tata Institute of Fundamental Research, Colaba, Mumbai, MH, India

Bhumika Mehta Department of Physics, Panjab University, Chandigarh, India

Dipak Kumar Mishra Bhabha Atomic Research Centre (BARC), NUCLEAR PHYSICS DIVISION (NPD), Trombay, Mumbai, MH, India

Jubin Mitra University of Oxford, Oxford, UK

B. Mohanty School of Physical Sciences, National Institute of Science Education and Research, Jatni, India;
Homi Bhabha National Institute, Training School Complex, Mumbai, India

Supratik Mukhopadhyay Applied Nuclear Physics Division, Saha Institute of Nuclear Physics, Kolkata, India;
Homi Bhabha National Institute, Mumbai, India;
Saha Institute of Nuclear Physics, A CI of Homi Bhabha National Institute, Kolkata, West Bengal, India

Tapan K. Nayak CERN, Geneva, Switzerland

Pawan Kumar Netrakanti Bhabha Atomic Research Centre (BARC), NUCLEAR PHYSICS DIVISION (NPD), Trombay, Mumbai, MH, India

Lalit Mohan Pant Bhabha Atomic Research Centre (BARC), NUCLEAR PHYSICS DIVISION (NPD), Trombay, Mumbai, MH, India

Pabitra Kumar Paul Department of Physics, Jadavpur University, Kolkata, India

Rajarshi Raut UGC-DAE Consortium for Scientific Research, Kolkata, India

Pradip Kumar Sahu Institute of Physics, Bhubaneswar, Odisha, India;
Homi Bhabha National Institute, Mumbai, India

Sanjib Kumar Sahu Institute of Physics, Bhubaneswar, Odisha, India;
Homi Bhabha National Institute, Mumbai, India

M. N. Saraf Department of High Energy Physics, Tata Institute of Fundamental Research, Colaba, Mumbai, MH, India

Sandip Sarkar Applied Nuclear Physics Division, Saha Institute of Nuclear Physics, Kolkata, India;
Homi Bhabha National Institute, Mumbai, India

B. Satyanarayana Department of High Energy Physics, Tata Institute of Fundamental Research, Colaba, Mumbai, MH, India

Raman Sehgal Bhabha Atomic Research Centre (BARC), NUCLEAR PHYSICS DIVISION (NPD), Trombay, Mumbai, MH, India

Prachi Sharma Department of Physics, Panjab University, Chandigarh, India

Tanvi Sheokand Department of Physics, Panjab University, Chandigarh, India

R. R. Shinde Department of High Energy Physics, Tata Institute of Fundamental Research, Colaba, Mumbai, MH, India

J. B. Singh Department of Physics, Panjab University, Chandigarh, India

Sagar Sonavane Department of High Energy Physics, Tata Institute of Fundamental Research, Colaba, Mumbai, MH, India

S. Thoi Thoi Department of High Energy Physics, Tata Institute of Fundamental Research, Colaba, Mumbai, MH, India

Dayong Wang School of Physics, Peking University, Beijing, China

Yue Wang School of Physics, Peking University, Beijing, China

Licheng Zhang School of Physics, Peking University, Beijing, China

Invited Papers

The Practice of Gamma-Ray Spectroscopy: Here and Now

Rajarshi Raut

1 Introduction

The many-body complex system called the nucleus exhibits myriad intriguing phenomena, particularly in its excited states. The practice of γ-ray spectroscopy aspires observation and decipherment of these many phenomena manifested by the excited nuclei. The tools and the techniques herein have extensively evolved over several decades to their current form that can now unravel even the rarest and the most exotic idiosyncrasies in the nuclear panorama. This article aspires to provide an overview on γ-ray spectroscopy through discussions on its essential components. The first is population of nuclei in their excited states, and de-excitations therefrom, that causes emission of radiations. The next component of the pursuit is the detection of the emitted γ-rays that is based on their interactions with the detector, chosen for the purpose. The subsequent step is to process the detection, obtained in the form of a pulse from the detector, for retrieving information of interest. And, finally, the acquired information needs to be converted into data that are stored for eventual processing and analysis in order to conclude on the physics being aspired in the relevant study.

The study of nuclear excitations commences with the production of the nuclei in their excited states, typically accomplished through nuclear reactions carried out at the accelerators. One of the most commonly used one is compound nucleus (CN) fusion-evaporation reactions. Herein, an energetic projectile, delivered by the accelerator, bombards on a static target nucleus and, depending on the beam energy and other factors (cross section etc.), fuses to form a compound system. The compound nucleus is an extremely energetic system and tends to de-excite, initially through emission of light particles, such as neutron, proton and alpha and eventually through

R. Raut (✉)
UGC-DAE Consortium for Scientific Research, Kolkata Centre, Kolkata 700106, India
e-mail: rajarshi.raut@gmail.com

R. N. Patra (ed.), *Advanced Radiation Detector and Instrumentation in Nuclear and Particle Physics*, Springer Proceedings in Physics 282,
https://doi.org/10.1007/978-3-031-19268-5_1

emission of γ-rays. The practice of γ-ray spectroscopy is essentially about detecting these γ-rays to infer on the level structure of the emitting nuclei. The details can be looked up in a range of texts such as by Regan [1] and Casten [2].

2 Detection

At the basis of detection of radiations is their interaction with the detection medium (matter). A γ-ray incident on a medium may or may not interact therein. If it does, there are different possible mechanisms with varying impact on the end-result of detection. The γ-ray can undergo photoelectric absorption or Compton scattering or, in case of high energy γ-rays $\geq$1.02 MeV, pair production. The photoelectric interaction dominates at the lowest energies while the pair production gains significance only at the highest ones. The Compton scattering remains the most probable mechanism of interaction for a wide range of incident γ-ray energies. As far as the spectroscopic objectives are concerned, of only pertinence is whether, as the result of one or more interactions, the full energy of the incident γ-ray is completely contained in the medium.

The solid state detectors, with higher matter density than the gas based ones and with increased probability of interaction, have emerged to be of preference in the practice of γ-ray spectroscopy. These detectors can be of scintillators or semiconductors [3, 4]. In γ-ray spectroscopy, some of the earliest measurements were based on NaI(Tl) and BaF_2 scintillators that are merited with superior detection efficiency. However, the scintillator detectors are plagued with poor energy resolution that come in the way of their usage for spectroscopic goals.

The advent of semiconductor based detectors revolutionized the practice of γ-ray spectroscopy through improved energy resolution, to typical values of ~2 keV at ~1 MeV, and has unleashed the era of modern spectroscopy wherein even the most complex excitation pattern of the nuclei could be deciphered. The semiconductor detectors are basically junction diodes of Si or Ge wherein the depletion region represents the active volume of the detector. Ge ($Z = 32$) is of preference in γ-ray detection, compared to the Si ($Z = 14$). The semiconductor junction detectors are operated under a reverse bias voltage effecting into the depletion region getting extended to almost the entire crystal volume. An interacting radiation create electron-hole pairs therein and these migrate under the existing (reverse bias) electric field towards their respective collector ends. The signal that results encodes information of interest on the detected radiation. The Ge material used for the detectors is High Purity Ge (HPGe), characterized with $\sim 10^{10}$ impurity atoms/cm^3 [3]. This facilitates an increased width of the depletion region for a given bias voltage and ensures maximizing the active volume of the detector. The Ge based detectors, however, need to be operated at liquid nitrogen temperature (77 K) in order to restrict the thermally generated leakage current.

A γ-ray spectrum is characterized by full-energy peaks and Compton background. The latter is required to be restricted for ensuring spectroscopic objectives. The

HPGe detectors are conventionally used with an Anti Compton Shield (ACS) for the purpose. The ACS is a scintillator based detector, typically using BGO (Bismuth Germanium Oxide) or NaI(Tl). The idea is to detect the Compton scattered photons, from the HPGe crystal, in the surrounding ACS and reject those events through a logic implementation during the pulse processing. The application of the ACS to reduce the Compton background is called an active shielding. The background also has a component from interaction of high energy γ-rays of cosmic origin, or other ambient sources, with the detector medium and efforts to reduce the same are based on what is called the passive shielding. This is implemented through housing the experimental setup in a canopy of high-Z material or by setting up the experiments in underground laboratories.

Apart from its efficiency and energy resolution, the timing characteristics of a detector are of extensive considerations particularly in the context of the settings in the pulse processing circuitry as well as its application in the investigation of nuclear level structure. The pulse resulting from a detector is characterized by a rise time, an (voltage) amplitude and a decay time. The amplitude depends on the number of charge carriers produced in the detector, following the interaction of a radiation therein and thus encodes the energy deposited in the process. The rise time reflects the time of collection of the charge carriers following their production in the radiation interaction. The rise time provide a marker to the so called "time of detection" that is applied for defining of timing correlations (coincidence, anti-coincidence) between different detectors in a setup and facilitates identification of the events of interest. For instance, the signals resulting from the HPGe detector and the corresponding ACS, following one Compton event that deposits energy in both, need to be correlated in time to be identified as resulting from the interaction of the same γ-ray in the two detectors (and thus to be rejected).

Larger detector volume would result into better efficiency and the same is implemented through composite HPGe detectors such as the clover. A HPGe clover has four HPGe crystals enclosed in a common cryostat. The packing of the crystals is close and indicates a realistic possibility that in the event of an incident γ-ray undergoing Compton scattering in one of the crystals, the scattered photon, carrying the residual energy, would go and interact with one or more of the adjacent crystals and cumulatively deposit the entire incident energy in the clover detector. This possibility translates into an increased efficiency, as was aspired for HPGe detectors, through what is called the addback mode of operating the clover. The clover detector is now a basic tool routinely applied for γ-ray spectroscopy measurements in this country as well as across the globe.

3 Processing

The next step towards arriving at the aspired physics is processing of the signal obtained from the detector. Such a signal is generally extracted from the preamplifier of the detector. The preamplifier is the first component of the pulse processing circuitry that is typically positioned very close to the detector. In fact for the HPGe detectors, the preamplifier is located inside the enclosure common to the crystals, albeit the latter are in vacuum while the preamplifier board is not, so as to facilitate easy access for maintenance. The preamplifier signal of a HPGe detector is characterized by a rise time of $\sim$ tens of ns, amplitude of $\sim$ tens of mV/MeV, and a decay time extending to $\sim$ tens of μs. The amplitude is proportional to the amount of charge carriers produced from a radiation interaction with the detector medium, and the energy deposited in the process. The preamplifier signal has to be processed for extracting information of primary interest on the detected radiation. These are the energy deposited by the radiation in the detector medium and the time of detection. For many decades, pulse processing has been implemented through use of modular electronics fabricated as per one of the global standards, such as the NIM (Nuclear Instrument Module) and the CAMAC (Computer Automated Measurement And Control).

The energy information from the preamplifier signal comes through its processing by a spectroscopy amplifier. The module is set to amplify the preamplifier pulse, through a gain factor, as well as filter/shape it for faithful extraction of pulse height, that represents the energy deposited in the corresponding detection. The pulse processing for extracting the timing information from the detector pulse output principally proceeds through application of modules such as a Timing Filter Amplifier (TFA) and a Constant Fraction Discriminator (CFD) to generate a time marker that is independent of the amplitude and the rise time of a detector pulse and represents the time of detection. The time marker facilitates identification of correlations between different detectors in the setup and that of the events of interest. For instance, the correlation between the HPGe and the corresponding ACS helps identifying the Compton events and implement the Compton suppression. The Compton suppressed time marker resulting from a Compton suppressed HPGe is typically used to define an event trigger in an array of such detectors. An event is the fulfillment of a condition subject to which the data is acquired. An event in an array of detectors typically corresponds to a user defined multiplicity of detectors firing in coincidence, within a time window, manifested in an overlap between their time markers. The pulse processing techniques elaborated in this section have been and are being routinely practiced in the country, both using commercially available modules as well as modules developed herein.

4 Data

The principal hardware for data acquisition is an ADC (Analog-to-Digital Converter) that reads-out the amplitude information of the amplifier pulse and facilitates recording the same as a number (channel number) in the data file. An ADC is characterized by a resolution, expressed in bits such as 12-bit, 13-bit, 14-bit, and indicates the maximum of the channel number it can generate. For instance, a 13-bit ADC can give out a maximum channel number of $2^{13} - 1 = 8191$. Apart from reading-out the amplitude information from the amplifier pulse, as a record of the energy deposited in the corresponding detector, the time of detection is also to be recorded in the data, particularly if an array of detectors is used for measurements. The Time-to-Digital Converter (TDC) is used for the purpose. The timing record is made from the time difference between a signal representing the time of detection in each detector, that triggers an event, with respect to a common signal and converting this difference into a channel number of the TDC. The commonly used format of the acquired data is the list mode format. Data in this format is an event-by-event record of the energy and the timing of the detections along with the identity of the detector elements, that constitute the event.

The advent of digital signal processing, and its applications in nuclear physics experiments, has revolutionized the methodology of processing and acquisition during the recent years. The technology, merited with fast processing and increased throughput, has facilitated acquisition of enhanced event rates that typically characterize the large array of detectors set up for γ-ray spectroscopy measurements. The Nuclear Physics group at the **UGC-DAE CSR**, has made significant contributions in the development and implementation of digitizer based pulse processing and data acquisition for nuclear structure studies in the country [5].

5 Data Reduction and Analysis

The first step in the data processing exercise is the energy calibration of the (ADC) channel numbers. The energy calibration is facilitated by the use of standard radioactive sources, such as ^{60}Co, ^{152}Eu, ^{133}Ba, ^{137}Cs, with known characteristic γ-ray energies. For an array of γ-ray detectors, the energy calibration is followed by what is called gain matching. It is necessary that the energy represented in a given channel number for a detector in the array is identical to the energy represented by the same channel number recorded for any other detector in the array. The unambiguous assignment of observed γ-ray transitions to a nucleus populated in an experiment typically follows γ-γ coincidence information. The analysis of data acquired in γ-ray spectroscopy measurements, that are targeted at probing the level structure of nuclei, proceeds through construction of varied kind of matrices, symmetric as well as asymmetric, so as to extract different measurables therefrom. These include, apart from identifying the coincidence relationships between the observed transitions, their angular correlations and the linear polarization of the γ-rays. The correlation and the polarization measurements facilitate in assigning the multipolarity and electromagnetic character of the transitions and thus the spin-parities of the associated levels. Another measurement of much significance is that of level lifetimes and the same is extracted through a variety of methods, depending on the lifetimes being addressed. Significant developments have recently been made in the relevant domain [6, 7] and have resulted in determination of level lifetimes with reduced uncertainties. Further details, on data analysis for γ-ray spectroscopy can be found in several papers and review articles, such as the ones by Regan [1] and Bhowmik [8].

6 The Indian National Gamma Array

The Indian National Gamma Array (INGA) has been the principal tool for pursuing nuclear structure studies, with in-beam γ-ray spectroscopy, in this country for around two decades now. It is an array of Compton suppressed HPGe clover detectors setup by a collaboration of Institutes and Universities across the nation. The participating organizations include **UGC-DAE CSR**, TIFR (Mumbai), SINP (Kolkata), VECC (Kolkata), IUAC (New Delhi), BARC (Mumbai) and several others. The user community for this facility is a widespread one and extends to institutions outside the country as well. The facility is set up at regular intervals in one of the accelerator centers in the country, VECC, IUAC and TIFR. The facility has evolved in its merits through the many experimental campaigns that have been hosted since its commencement. The first setup was at the Pelletron LINAC Facility (then TIFR-BARC Pelletron) at TIFR, in around 1999–2000 and consisted of 8–9 Compton suppressed clover detectors. The subsequent campaign at IUAC (then NSC), around 2002–2003, was again based on 8 Compton suppressed clovers as well as certain ancillary detectors for charged particles, neutrons, and recoils that facilitated improved identification of the reaction channel of interest. These campaigns of INGA were largely based on commercially available pulse processing electronics of NIM standards and CAMAC based data acquisition hardware, albeit the acquisition software were developed at BARC and IUAC, respectively, for the first and the second programmes. The acquisition program used in the first campaign was AMPS, developed at BARC, while the one used during the second campaign, at IUAC, was CANDLE. This was followed by a significant development at IUAC wherein compact clover electronics modules were designed and fabricated such that each such module could support the pulse processing of one clover along with its ACS. These modules were put to use in the INGA campaign at VECC cyclotron, during 2004–06, with 8–10 Compton suppressed clovers [9]. The data acquisition used in this campaign was CAMAC-based and running on the LAMPS [10] software, developed at BARC as an updated linux version of the previous AMPS code. A major augmentation in the number of detectors followed, and the next INGA campaign at IUAC, during 2008–09 [11], had around 22–24 Compton suppressed clover detectors in the setup. The compact electronics modules, which had been developed earlier, significantly helped in sustaining the campaign. The data acquisition hardware, apart from the commercially available modules, also consisted of ADCs fabricated at IUAC and was supported by an updated version of CANDLE. The next major stride in the history of INGA was taken during its campaign at TIFR, during 2009–13, with around 24 Compton suppressed clover detectors. Herein, a digitizer based data acquisition system was used as a maiden instance in the practice of γ-ray spectroscopy in the country [12]. The use of digital signal processing manifolded the merits of the facility that could now handle increased event rates characterizing the large array of detectors. The following and the most recent campaigns of INGA have progressed simultaneously at both IUAC and VECC. The one at IUAC has around 16 Compton suppressed clovers. It is largely based on the same infrastructure of the previous campaign hosted therein,

Fig. 1 The detector setups during different INGA campaigns. In the upper panel, from left to right, is the first INGA st TIFR, second campaign at IUAC, third campaign at VECC, and following campaign at IUAC. The lower panel shows the setups at two recent campaigns, at TIFR on the left and at VECC on the right

back in 2008, but with an improved data acquisition system, running on CANDLE, and facilitated with capabilities for handling higher event rates. The campaign at VECC has been based on around 8 Compton suppressed clover detectors and upto 2 LEPS, for detecting x-rays and low energy γ-rays. Most importantly, this experimental programme at VECC has been supported by a digitizer based pulse processing and data acquisition system that is conceptualized and implemented by **UGC-DAE CSR** [5], as one of the major developments in the progress of INGA and has been effectively used in over 30 experiments at VECC. That was a very short history of the INGA facility through its evolution into the current being. Figure 1 shows the photographs of the detector setups in the different campaigns. Efforts are in progress to better the facility, to increase its usage, and to explore the current frontiers of nuclear structure research by exercising the many developments that have been accomplished in the process.

The help and support of the INGA Collaboration is gratefully acknowledged. The INGA Project is partially supported by the Department of Science and Technology, Government of India (Grant No. IR/S2/PF-03/2003-III).

References

1. P. Regan, *Post Graduate Nuclear Experimental Techniques Course Notes* (University of Surrey, Guildford, 2003)
2. R.F. Casten, *Nuclear Structure from a Simple Perspective*, 2nd edn. (Oxford Science Publications, Oxford, 1999)
3. G.F. Knoll, *Radiation Detection and Measurement* (Wiley, New York, 1999)
4. W.R. Leo, *Techniques for Nuclear and Particle Physics Experiments* (Springer, Berlin, 1994)
5. S. Das, S. Samanta, R. Banik, R. Bhattacharjee, K. Basu, R. Raut, S.S. Ghugre, A.K. Sinha, S. Bhattacharya, S. Imran, G. Mukherjee, S. Bhattacharyya, A. Goswami, R. Palit, H. Tan, Nucl. Instr. Meth. Phys. Res. A **893**, 138 (2018)
6. R. Bhattacharjee, S.S. Bhattacharjee, K. Basu, P.V. Rajesh, R. Raut, S.S. Ghugre, D. Das, A.K. Sinha, L. Chaturvedi, U. Garg, S. Ray, Phys. Rev. C **90**, 044319 (2014)
7. S. Das, S. Samanta, R. Bhattacharjee, R. Raut, S.S. Ghugre, A.K. Sinha, U. Garg, R. Chakrabarti, S. Mukhopadhyay, A. Dhal, M.K. Raju, N. Madhavan, S. Muralithar, R.P. Singh, K. Suryanarayana, P.V. Madhusudhana Rao, R. Palit, S. Saha, J. Sethi, Nucl. Instr. Meth. Phys. Res. A **841**, 17 (2017)
8. R.K. Bhowmik, in *Structure of Atomic Nuclei*, ed. by L. Satpathy (Narosa Publishing House, Chennai, 1999)
9. R. Raut et al., Proc. DAE. Symp. Nucl. Phys **47B**, 578 (2004)
10. https://www.tifr.res.in/~pell/lamps.html
11. S. Muralithar, K. Rani, R. Kumar, R.P. Singh, J.J. Das, J. Gehlot, K.S. Golda, A. Jhingan, N. Madhavan, S. Nath, P. Sugathan, Nucl. Instr. Meth. Phys. Res. A **622**, 281 (2010)
12. R. Palit, S. Saha, J. Sethi, T. Trivedi, S. Sharma, B.S. Naidu, S. Jadhav, R. Donthi, P.B. Chavan, H. Tan, W. Hennig, Nucl. Instr. Meth. Phys. Res. A **680**, 90 (2012)

Applications of Radiation Detectors to Society

Aafke Kraan

1 Introduction

The main challenge in radiotherapy treatments is how to deliver high dose to the tumor region, while minimizing the irradiation of healthy tissue. An important modality that has gained significant interest in the past decades is irradiation with charged ions (protons or carbon ions). The characteristic dose deposition profile (Bragg peak) of charged particles allows to deposit dose much more locally than the photons, so dose in healthy tissue can be minimized [1, 2]. By the end of 2020, more than 290000 patients had been treated worldwide with particle therapy, of which about 250000 with protons, about 40000 with carbon ions, 3500 with helium, pions and with other ions [3]. In India, the Apollo Proton Cancer Centre was inaugurated on January 25, 2019 in Chennai. This facility represents South East Asia's first Proton Therapy Centre.

The high selectively of particle beams also implies that treatments with charged particles are more sensitive to uncertainties than photon treatments, because of their steep dose profile [4]. Among the main error sources are setup errors and anatomical modifications in patients during the treatment course and range errors due to uncertainties in the conversion from Hounsfield units to particle stopping [4]. In the last decades, various imaging modalities have been developed to ensure that particle therapy treatments are planned and delivered correctly (see for instance [5]). One of the methods is *in-vivo* non-invasive range monitoring, which is based on the detection of fragments from nuclear physics reaction processes of the incoming particle beam with the tissue of the patient [6–8]. Here no additional dose is delivered to the patient ('non-invasive') in addition to the planned therapeutic dose. Moreover the information can be obtained in real-time with dedicated equipment, while the

A. Kraan (✉)
INFN Pisa, Largo Bruno Pontecorvo, 3/Edificio C 56127 Pisa, Italy
e-mail: aafke.kraan@pi.infn.it

R. N. Patra (ed.), *Advanced Radiation Detector and Instrumentation in Nuclear and Particle Physics*, Springer Proceedings in Physics 282,
https://doi.org/10.1007/978-3-031-19268-5_2

patient is being treated. This minimizes the effect of organ motion and physical and biological washout (see Sect. 3.1).

In this lecture I will briefly summarize the physics of these non-invasive methods. Moreover, focusing on in-beam PET monitoring, various ongoing research developments will be highlighted.

2 Non-invasive Treatment Monitoring

When a particle beam passes through patient tissue, nuclear interactions occur between the incoming particle and the nuclei of the tissue of the patient (see Fig. 1). Although these processes are generally considered undesired (they modify the physical and biological dose), they can be used to estimate the beam range in tissue, if they can be detected with dedicated detector systems. Let's discuss in more detail how these nuclear reactions give rise to different types of secondary particles.

A nuclear interaction can be described as a sequence of three stages [9, 10], each of which can give rise to different types of secondaries. The fastest stage of a nuclear reaction (within a time $t < 10^{-22}$) consists of one or more interactions between the nucleons of the projectile and target nuclei. These interactions can give rise to high-energy secondary fragments, predominantly produced in the forward directions, with energies that may be close to the projectile energy. Then a pre-equilibrium stage ($t \sim 10^{-20}$) follows, in which other, lower energy, secondary particles are produced (mainly nucleons). The final latest stage of the nuclear reaction ($t \sim 10^{-16}$) involves evaporation, fragmentation, fission and finally de excitation. Here secondaries are produced with lower energies, that are emitted more or less isotropically in the rest frame of the mother nucleus (target or projectile).

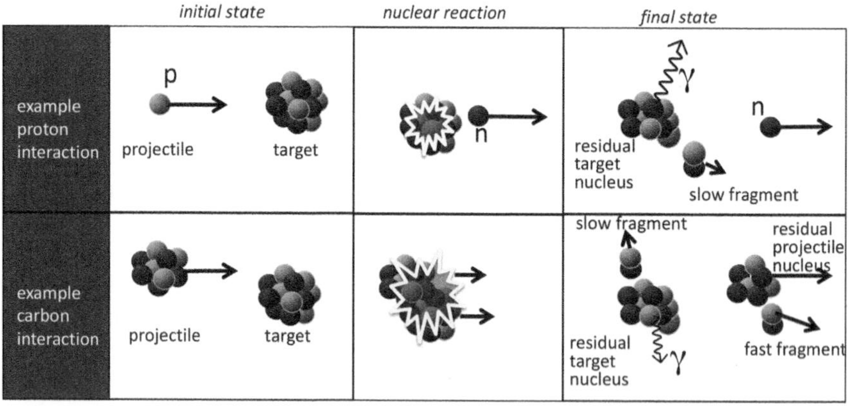

Fig. 1 Example of a proton and a carbon interaction, with various nuclear reaction products produced

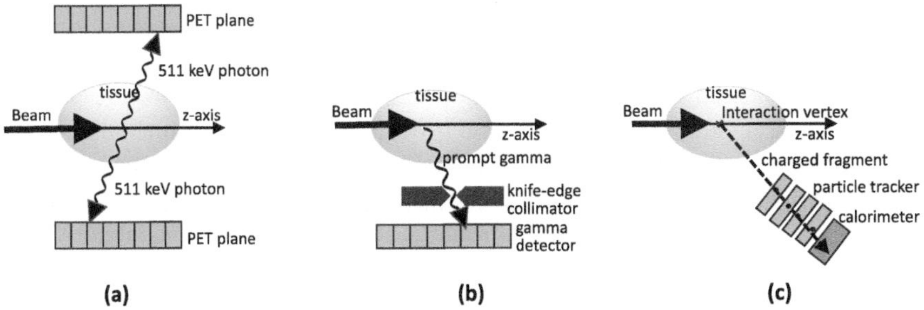

Fig. 2 Concept of treatment monitoring in particle therapy with PET imaging (**a**), prompt gamma imaging, here with knife-edge collimator (**b**), and charged fragment detection (**c**)

It must be noted that nuclear reactions cannot be neglected from a dose point-of-view. For instance, for 160 MeV protons hitting a water target, roughly 20% of the incident protons will not reach the Bragg peak position at about 16 cm. For 290 MeV/u carbon ions impinging on water, as much as 50% of the ions have undergone a nuclear reaction at the end of range. The fluence loss due to primary interactions are generally included in the treatment planning systems for the correct calculation of physical and biological dose. Of all the nuclear fragments produced, only a small fraction can be used for range monitoring.

The three modalities discussed in this lecture are PET imaging, prompt gamma imaging, and charged fragment detection. They are sketched in Fig. 2.

3 PET Imaging for Particle Treatment Monitoring

Already in 1982, Chatterjee recognized that β^+ emitting nuclei were produced during irradiation with therapeutic particle beams, and measured the two coincidence photons using a positron-emitting beam analyzer system consisting of two banks of NaI(Tl) crystals [11]. A large variety of fragments is produced in human tissue during irradiation, including β^+ emitting fragments like ^{15}O, ^{11}C, ^{10}C, ^{12}N, ^{13}N, and so on [12, 13]. Figure 2a illustrates the concept of PET imaging in particle therapy. This β^+ nuclei decays at some time after production, depending on the value of their half-life. A positron is emitted, which travels a small distance (of the order mm) and then annihilates with an electron in the medium into two coincidence photons of 511 keV. These photons can be measured with a PET detector. The PET activity profile is indirectly correlated with the dose profile. Much research has been dedicated to range monitoring with PET in the last two decades, summarized for instance in various reviews [6, 8, 13, 14].

Different PET data acquisition strategies have been applied, which can be divided in offline and online data acquisition. In the offline technique, data acquisition starts typically minutes after beam delivery and outside the treatment room ('offline'), leading to a loss of signal of short-lived isotopes. Other problems include organ motion, co-registration problems, and physiological washout, making this technique generally considered sub-optimal [14].

In online techniques, data are acquired inside the treatment room ('online'). We can distinguish data acquisition during or right after dose delivery without moving the patient ('in-beam'), and data acquisition after beam delivery where the patient is moved to an in-room PET scanner ('in-room'). The advantage of data taking during dose delivery is that it allows to detect short-lived isotopes, that would otherwise be lost. Moreover effects like wash-out and patient movements are much smaller than in offline PET (see for instance [15, 16]). Most recent efforts in PET imaging as treatment monitoring technique are therefore ongoing in this direction. Below we highlight some ongoing research developments in this context.

3.1 In-Beam PET Monitoring Systems

Regarding in-beam PET imaging, several dedicated PET systems were developed in the past, all facing the challenge of making a PET system compatible with the beam delivery system.

At the heavy ion medical accelerator in Chiba (HIMAC), various geometries including the cylindrical OPENPET geometry [17] displayed in Fig. 3, that was also tested in clinical conditions [18].

The scintillators for photon detection are Zr-doped GSO (GSOZ) scintillators with a crystal size of $2.8 \times 2.8 \times 7.5$ mm^3, connected to position sensitive photomultiplier tubes. The spatial resolution and sensitivity were 2.6 mm and 5.1% for the open mode.

Another type of geometry used for in-beam PET data acquisition is the planar geometry, where two flat PET planes are used to detect the β^+ activity. This was the configuration used in the oldest in-beam PET monitoring system, used at GSI of patient treatments with monitoring in-beam PET monitoring of patients treated

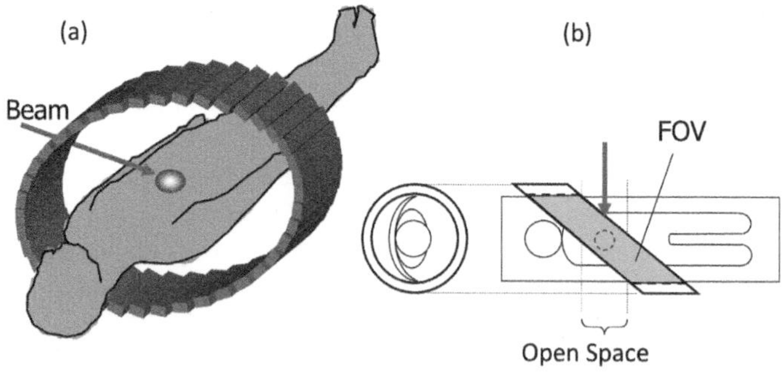

Fig. 3 Conceptual illustrations of OPENPET geometry with an accessible open space to the patient for beam delivery. The red arrow indicates the beam direction. Taken from [18]

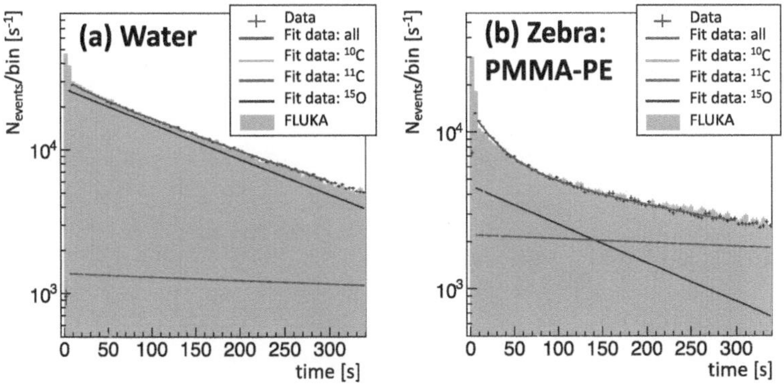

Fig. 4 For two different phantoms the decay rate as a function of time for the data (red) and Monte Carlo simulations (filled blue area), with for the latter the fitted contributions from ^{15}O (blue), ^{11}C (dark grey), and ^{10}C (light grey)

with ^{12}C ions [19, 20]. Planar configurations were also applied in the past in Japan at Chiba [21] and at the Kashiwa carbon ion treatment facility [22]. Other past applications featuring compact movable planar PET systems were been developed at INFN, including the DOPET system tested at CATANA and CNAO [23, 24] in Italy.

Recent and ongoing research is being done with a larger version of the DOPET planar system, 15 × 15 cm^2. It was tested at the Cyclotron Centre of the Bronowice proton therapy centre in Krakow, Poland. Not only spatial distributions were investigated, but also time profiles. In fact, the time decay curve of the activated material can provide independent additional information about the treatment and possible changes. This is because the shape of the time profile curves depends on the decaying isotopes and thus on the composition of the irradiated tissue [25–27].

This is demonstrated in Fig. 4, where the 1-D decay rate is displayed as a function of time in two different materials. In both figures we have shown the fitted contributions from ^{15}O ($t_{1/2} = 2$ min), ^{11}C ($t_{1/2} = 20$ min), and ^{10}C ($t_{1/2} = 19$ s). These kind of evaluations can also be done voxel-by-voxel. If variations in the fitted decay rate curves are observed, this can be an indication that changes have occurred in the patient.

Another recent development in in-beam PET imaging concerns the bi-model in-beam treatment monitoring system INSIDE [28]. This system combines an in-beam PET system and a secondary charged particle tracker, called the INSIDE system, is in a clinical trial phase at CNAO. The PET system has two planar heads that are spaced 60 cm apart, each with an active area of 10 × 25 cm^2. Lutetium Fine Silicate (LFS) is used as scintillating material, with 3.1 × 3.1 × 20 mm^3 crystals and 3.2 mm pitch. The readout electronics consists of a 16 × 16 array of Hamamatsu Silicon Photomultiplier (SiPM) coupled one-to-one to each crystal. At present the system is undergoing a clinical trial at CNAO, focusing on head-and-neck cases [29]. Much effort is currently ongoing to develop appropriate analysis methods [30, 31]. A new

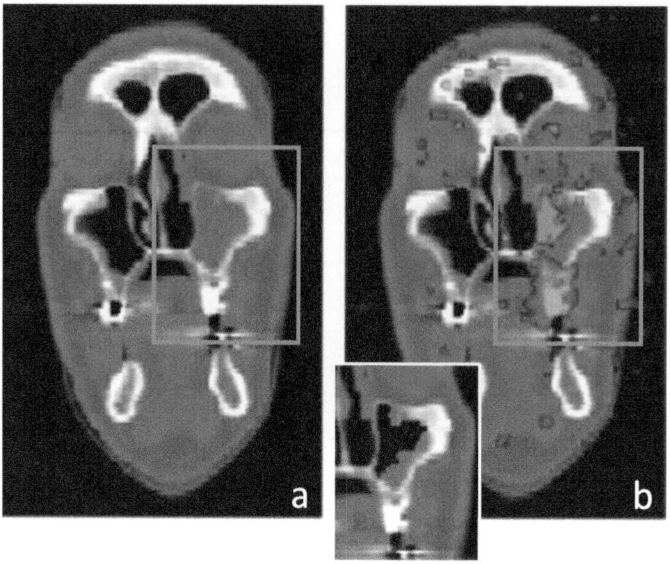

Fig. 5 p-value maps for various sizes of emptied volumes (FLUKA Monte Carlo simulations) superimposed on the planning CT scan. The green and red regions represent zones with a statistically significantly smaller and larger activity than expected, respectively. The small insets in the slices show the corresponding region that is changed

development in this direction is the voxel-based morphometry approach [32] to detect anomalous patterns. The goal here is to create maps with p-values, that indicate the probability that a morphological change occurred somewhere. An example of such a map, here based on FLUKA Monte Carlo simulations [33], is given in Fig. 5. The green and red regions represent zones with a statistically significantly smaller and larger activity than expected, respectively, strongly suggesting a morphological change.

Finally, much research is ongoing to detect short-lived PET isotopes [34, 35]. The potential of imaging ^{12}N for range verification was investigated experimentally for proton irradiation with the experimental KVI cyclotron accelerator in Groningen, the Netherlands. Millimeter precision range measurements were obtained when imaging the ^{12}N emitters in both graphite and PMMA targets with proton pencil beams [35]. Recently, PET monitoring of ^{12}N emitters was also investigated in Helium irradiation [36]. One of the main challenges here is the implementation in clinical irradiation conditions.

4 Other Non-invasive Methods Based on Secondary Particle Detection

Nuclear reactions lead to other reaction products besides β^+ emitters.

- Promptly emitted photons, from the nuclear de-excitation phase of a nuclear reaction, with energies ranging from 0 to about 10 MeV. There is a correlation between the longitudinal prompt gamma profile and the dose profile, making it possible to exploit prompt gammas to retrieve the Bragg peak position. For comprehensive reviews we refer to recent works by Krimmer et al. [37] and Wroska et al. [38] and citations therein. The most common approach to do this is with a collimated gamma camera. By placing the camera at 90° with respect to the beam-axis and moving the device parallel to the beam-axis, a 1-D prompt gamma profile can be measured (see Fig. 2b).
- Interaction vertex imaging (IVI) for carbon ion-therapy range monitoring (see for instance [39–44]. The method is based on the detection of secondary charged particles with high enough energies that they can exit the patient during particle irradiation. The position profile of the reconstructed nuclear emission vertex is indirectly correlated with the beam range. (see Fig. 2c). Recently, this method has been tested for the first time to monitor patients enrolled in the INSIDE clinical trial at CNAO [44].

5 Conclusion

Non-invasive treatment monitoring is one of the modalities for in-vivo dose verification. In this lecture I described the physics and a some ongoing developments in this field.

Acknowledgements First of all, many thanks to the Organizing Committee of this workshop and in particular Rajendra Nath Patra for having invited me to give this lecture. I also thank the audience for the positive feedback. Finally, thanks to Alberto Del Guerra, Maria Giuseppina Bisogni and Giuseppe Battistoni for their comments on the presentation and proceedings.

References

1. R.R. Wilson, Radiological use of the fast protons. Radiology **47**, 487–91 (1946)
2. J.S. Loeffler, M. Durante, Charged particle therapy - optimization, challenges and future directions. Nat. Rev. Clin. Oncol. **10**, 411–424 (2013)
3. https://www.ptcog.ch/index.php/ptcog-patient-statistics
4. H. Paganetti, Range uncertainties in proton therapy and the role of Monte Carlo simulations. Phys. Med. Biol. **57**, R99–R117 (2012)
5. J. Seco, M.F. Spadea, Imaging in particle therapy: state of the art and future perspective. Acta Oncologica **543**(9), 1254–8 (2015)

6. A.C. Knopf, A. Lomax, In vivo proton range verification: a review. Phys. Med. Biol. **58**(15), R131-160 (2013)
7. M. Durante, H. Paganetti, Nuclear physics in particle therapy: a review. Rep. Prog. Phys. **79**(9), 096702 (2016)
8. A.C. Kraan, Range verification methods in particle therapy: underlying physics and Monte Carlo modeling. Front. Oncol. **5**, 150 (2015)
9. S.G. Mashnik, K.K. Gudima, R.E. Prael, et al., CEM03.03 and LAQGSM03.03: Event Generators for the MCNP6, MCNPX and MARS15 Transport Codes (2008), *Proceed. Joint ICTP-IAEA Adv, Workshop on Model Codes for Spallation Reactions*
10. A. Ferrari, P.R. Sala, The physics of high energy reaction, in *Proceed. Workshop on Nucl. Reaction Data and Nuclear Reactor Physics, Design and Safety ICTP*, vol. 2 (1998), pp. 424–532
11. A. Chatterjee, W. Saunders, E.I. Alpen, J. Alonso, J. Scherer, J. Llacer, Physical measurements with high-energy radioactive beams. Radiat. Res. **92**(2), 230–44 (1982)
12. W. Enghardt, P. Crespo, F. Fiedler, R. Hinz, K. Parodi, J. Pawelke et al., Charged hadron tumour therapy monitoring by means of PET. Nucl. Instr. Meth. A **525**, 284–88 (2004)
13. X. Zhu, G. El Fahkri, Proton therapy verification with PET imaging. Theranostics **3**(10), 731–40 (2013)
14. K. Parodi, J.C. Polf, In vivo range verification in particle therapy. Med. Phys. **45**, e1036 (2018)
15. S. Helmbrecht, W. Enghardt, K. Parodi et al., Analysis of metabolic washout of positron emitters produced during carbon ion head and neck radiotherapy. Med. Phys. **40**(9), 091918 (2013)
16. J.M. Pakela, A. Knopf, L. Dong, A. Rucinski, W. Zou, Management of motion and anatomical variations in charged particle therapy: past, present, and into the future. Front. Oncol. (2022)
17. H. Tashima, T. Yamaya, E. Yoshida, S. Kinouchi, M. Watanabe, E. Tanaka, A single-ring OpenPET enabling PET imaging during radiotherapy. Phys. Med. Biol. **57**(14), 4705–18 (2012)
18. T. Yamaya, OpenPET: a novel open-type PET system for 3D dose verification in particle therapy. J. Phys.: Conf. Series **777**, 012023 (2017)
19. W. Enghardt, K. Parodi, P. Crespo et al., Dose quantification from in-beam positron emission tomography. Radiother. Oncol. **73**(Suppl. 2), S96-8 (2004)
20. P. Crespo, G. Shakirin, W. Enghardt, On the detector arrangement for in-beam PET for hadron therapy monitoring. Phys. Med. Biol. **51**(9), 2143–63 (2006)
21. Y. Iseki, H. Mizuno, Y. Futami, T. Tomitani, T. Kanai, M. Kanazawa et al., Positron camera for range verification of heavy ion radiotherapy. Nucl. Instr. Meth. A **515**(3), 840–49 (2003)
22. T. Nishio, A. Miyatake, T. Ogino, K. Nakagawa, N. Saijo, H. Esumi, The development and clinical use of a beam ON-LINE PET system mounted on a rotating gantry port in proton therapy Int. J. Radiat. Oncol. Biol. Phys. **76**(1), 277–86 (2010)
23. G. Sportelli, N. Belcari, N. Camarlinghi, G.A. Cirrone, G. Cuttone, S. Feretti et al., First full-beam PET acquisitions in proton therapy with a modular dual-head dedicated system. Phys. Med. Biol. **59**(1), 43–60 (2014)
24. A. Topi, S. Muraro, G. Battistoni, N. Belcari, M.G. Bisogni, N. Camarlinghi, Monitoring proton therapy through in-beam PET: an experimental phantom study. IEEE Trans. Rad. Plas. Med. Sci. **4**(2), 194–201 (2020)
25. K. Grogg, N.M. Alpert, X. Zhu et al., Mapping ^{15}O product ion rate for proton therapy verification. Int. J. Rad. Onc. Biol. Phys **92**(2), 453 (2015)
26. J. Cho, G. Ibbott, M. Gillin et al., Determination of elemental tissue composition following proton treatment using positron emission tomography. Phys. Med. Biol. **58**(11), 3815–3835 (2013)
27. A. Kraan, S. Muraro, G. Battistoni, N. Belcari, G.M. Bisogni, N. Camarlinghi et al., Analysis of time-profiles with in-beam PET monitoring in charged particle therapy. J. Instrum. **14**, C02001 (2018)
28. G.M. Bisogni, A. Attili, G. Battistoni, N. Belcari, N. Camarlinghi, P. Cerello et al., INSIDE in-beam positron emission tomography system for particle range monitoring in hadrontherapy. J. Med. Im. **4**, 011005 (2016)

29. ClinicalTrials.gov Identifier: NCT03662373, INSIDE clinical trial
30. E. Fiorina, V. Ferrero, G. Baroni, G. Battistoni, N. Belcari, N. Camarlinghi, et al., Detection of interfractional morphological changes in proton therapy: a simulation and In Vivo study with the INSIDE in-beam PET. Front. Phys. 8 (2021)
31. A.C. Kraan, A. Berti, A. Retico et al., Localization of anatomical changes in patients during proton therapy with in-beam PET monitoring: a voxel-based morphometry approach exploiting Monte Carlo simulations. Med. Phys. **49**(1), 23–40 (2022)
32. K.J. Friston, A.P. Holmes, K.J. Worsley, J.P. Poline, C.D. Frith, R.S.J. Frack-owiak, Statistical parametric maps in functional imaging: a general linear approach. Hum. Brain Mapp. **2**, 189–210 (1994)
33. G. Battistoni, J. Bauer, F.T. Boehlen et al., The FLUKA code: an accurate simulation tool for particle therapy. Front. Oncol. (2016)
34. H.J.T. Buitenhuis, F. Diblen, K.W. Brzezinski, S. Brandenburg, P. Dendooven, Beam-on imaging of short-lived positron emitters during proton therapy. Phys. Med. Biol. **62**, 4654–72 (2017). https://doi.org/10.1088/1361-6560/aa6b8c
35. I. Ozoemelam, E. van der Graaf, M.-J. van Goethem, M. Kapusta, N. Zhang, S. Brandenburg et al., Feasibility of quasi-prompt PET-based range verification in proton therapy. Phys. Med. Biol. **65**(24), 245013 (2020)
36. I. Ozoemelam, E. van der Graaf, M.-J. van Goethem, M. Kapusta, N. Zhang, S. Brandenburg, P. Dendooven, Real-time PET imaging for range verification of helium radiotherapy. Front. Phys. (2020)
37. J. Krimmer, D. Dauvergne, J.M. Létang, É. Testa, Prompt-gamma monitoring in hadrontherapy: a review. Nuc. Instr. Meth. Phys. Res. A **878**, 58–73 (2018)
38. A. Wroska, Prompt gamma imaging in proton therapy - status, challenges and developments. J. Phys. Conf. Ser. **1561**, 012021 (2020)
39. U. Amaldi, W. Hajdas, S. Iliescu, N. Malakhov, J. Samaratti, F. Sauli et al., Advanced quality assurance for CNAO. Nucl. Instr. Meth. Phys. Res. A **617**, 248–9 (2010)
40. P. Henriquet, E. Testa, M. Chevallier, D. Dauvergne, G. Dedes, N. Freud et al., Interaction vertex imaging (IVI) for carbon ion therapy monitoring: a feasibility study. Phys. Med. Biol. **57**(14), 4655–69 (2012)
41. K. Gowsch, B. Hartmann, J. Jakubek, C. Granja, P. Soukup, O. Jäkel et al., Non-invasive monotoring of therapeutic carbon ion beams in a homogeneous phantom by tracking pf secondary ions. Phys. Med. Biol. **58**(11), 3755–73 (2013)
42. L. Piersanti, F. Bellini, F. Bini, F. Collamati, E. DeLucia, M. Durante et al., Measurement of charged particle yields from PMMA irradiated by a 220 MeV/u 12C beam. Phys. Med. Biol. **59**(7), 1857–72 (2014)
43. C. Finck, Y. Karakaya, V. Reithinger, R. Rescigno, J. Baudot, J. Constanzo et al., Study for online range monitoring with the interaction vertex imaging method. Phys. Med. Biol. **62**(24), 9220–9239 (2017)
44. M. Toppi, G. Baroni, G. Battistoni, M.G. Bisogni, P. Cerello, M. Ciocca, et al., Monitoring carbon ion beams transverse position detecting charged secondary fragments: results from patient treatment performed at CNAO. Front. Oncol. 11601784 (2021)

From Detector Simulation to Data Analysis in High Energy Physics Experiments

Shyam Kumar

1 Basic Concept of a Charged Particle Detector

When a charged particle passes through the detector, it creates electron-hole pairs (in semiconductors) or electron-ion pairs (for gaseous medium), which drift towards the electrode and create an induced signal on the electrode. Therefore, an incoming charged particle can be detected by the signal induced on the electrode. To identify the type of particle, we need to understand the mechanism of energy loss given by the Bethe-Bloch formula [4]. The energy loss as a function of momentum is plotted in the right panel of Fig. 1 for different particles in silicon. The signal detected by loss of 150 KeV of incident energy can be attributed to one of the particle shown by dotted horizontal line in Fig. 1. This implies that the specific energy loss is not enough to determine the type of incident particle (e.g. Electron, Muon, Pion, etc.). Therefore, to determine the type of particle, momentum is also required. The determination of momentum requires the reconstruction of charged particle trajectory under the magnetic field.

2 Determination of Momentum

Charged particles in a uniform magnetic field (B) experience the Lorentz force and hence the trajectory will be a helix as shown in the left panel of Fig. 2. In Cylindrical coordinates (R, ϕ, Z), it will be a circle in the R$-\phi$ plane and a straight line in the R-Z plane. We can determine the radius (R) of curvature by fitting the circle in the

S. Kumar (✉)
University and INFN, Bari, Italy
e-mail: shyam.kumar@ba.infn.it

R. N. Patra (ed.), *Advanced Radiation Detector and Instrumentation in Nuclear and Particle Physics*, Springer Proceedings in Physics 282,
https://doi.org/10.1007/978-3-031-19268-5_3

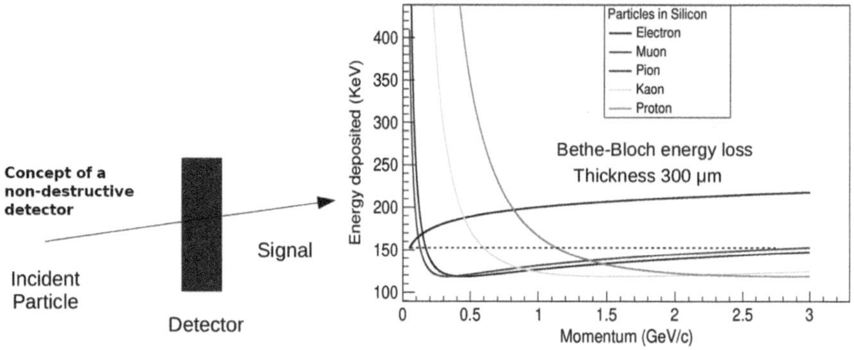

Fig. 1 (Left) Charged particle passing through the detector and (Right) energy loss for different charged particles inside silicon of 300 μm thickness plotted using the Bethe-Bloch formula

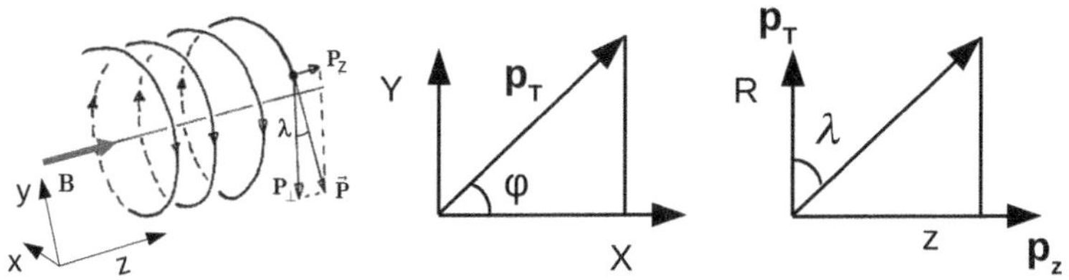

Fig. 2 (Left) Charged particle in the uniform magnetic field [6] and (Right) X-Y and R-Z plane showing azimuthal angle (ϕ) and dip-angle (λ)

R$-\phi$ plane and then the transverse momentum (p_{T}) can be determined by using the expression below.

$$p_{\text{T}}\ (\text{GeV/c}) = 0.3\ \text{B[T]}\ \text{R[m]} \tag{1}$$

We also need to fit the trajectory with the straight line in the R-Z plane, this can determine the dip-angle λ (angle between transverse momentum and total momentum direction), which will ultimately determine the z-component of momentum ($\text{p}_{\text{z}} = p_{\text{T}}\ \text{Tan}\lambda$) as shown in the right panel of Fig. 2. Since the particle creates the hit points on the detector plane, we can use these points to fit the particle trajectory (track). The final goal is the evaluation of both spatial coordinate and momentum components along the trajectory of the particle, and at the production vertex in particular, along with the uncertainties on these quantities [5].

3 Why Detector Simulation?

The performance of the HEP experimental apparatus (detector) is evaluated using numerical MC simulations. To understand the necessity of detector simulation, let's consider an example consisting of six silicon layers ($\sigma_{r\phi} = \sigma_z = 10\ \mu\text{m}/\sqrt{12}$) of

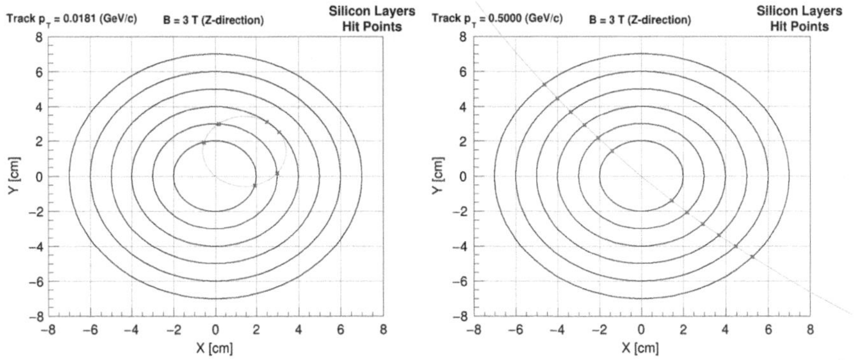

Fig. 3 (Left) Minimum $p_{\rm T}$ of track hitting three silicon layers (Right) Track of $p_{\rm T}$ = 0.5 GeV/c giving hits in the silicon layers

thickness 50 μm as shown in Fig. 3. To determine the radius of curvature of the track requires at least three points (i.e. at least three detector layers). This ultimately determines minimum transverse momentum reconstructed ($p_{\rm T_{min}} = 0.3 \times 3 \times 0.02 = 0.018$ GeV/c) with the experimental setup in the left panel of Fig. 3. A track of $p_{\rm T} =$ 0.50 GeV/c giving hits in all-silicon layers is also shown in the right panel of Fig. 3. Obvious questions are: how many layers do we need? What is the best distance between layers? What are the uncertainty on the measured points and momentum measurement? Precise determination can only be obtained by the detector simulation.

4 Concept of Detector Simulation

We first simulate a charged particle of given (known) momentum in our experimental setup (see Fig. 4). The intersection of the track with the detector planes creates the true points. In the next step, we try to reconstruct these points using the signal on detectors and use them to determine the reconstructed momentum. In fact, there is a difference between true and reconstructed position and true and reconstructed momentum because of finite uncertainty associated with the size of pixels (measuring points), the effect of multiple scattering in the material, residual bias in the fitting algorithm of tracks, etc. Therefore, one evaluates the RMS of difference of reconstructed and true position or momentum also known as spatial and momentum resolutions respectively. These are very crucial parameters that drive the physics performance, and that should optimize in the design of the apparatus. Simulations are used with this purpose, and they include a particle generator, a transport code for Geometry and Tracking [2].

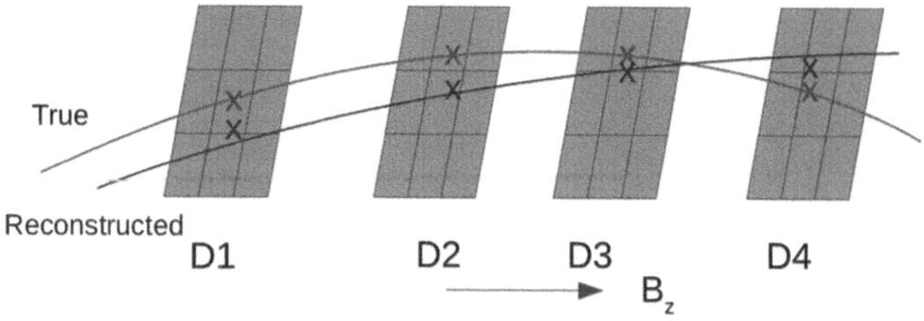

Fig. 4 Detector layers (D1, D2, D3, D4) together with true track (red line), true points (red markers), reconstructed track (blue line), and reconstructed hits (blue markers)

4.1 CEE TPC Simulation

CSR External Target Experiment (CEE) experiment [3] is a future facility to be built at the IMP, CAS, China. The detector setup (beam in Z-direction) consists of a tracking system based on a Time Projection Chamber (TPC) that exploits Gas Electron Multiplier (GEM) detectors as shown in the left panel of Fig. 5. Simulation of TPC is done in the CeeRoot framework based on the FAIRROOT [1]. When a charged particle passes through the TPC, it creates ionization (electron-ion pairs) which drift and diffuse till the first GEM plane, makes an avalanche in the GEM layers then creates the induced signal on the pads. X and Z positions can be reconstructed by the signal on the pads, while Y-coordinate is reconstructed from the drift time as shown in the right panel of Fig. 5. The pad plane consisting rectangular pads of size 5×12 mm^2 (X-Z direction) used in the simulation. Protons are simulated (uniform in θ : [20, 60] & ϕ : [0, 2π]) in the momentum range 0.10–1.50 GeV/c using BoxGenerator in the CeeRoot framework. The transport of protons inside the detector material is done by using a particle gun (G4ParticleGun) in GEANT4. Since TPC has a gaseous medium, we considered the Monte Carlo (MC) points (or true

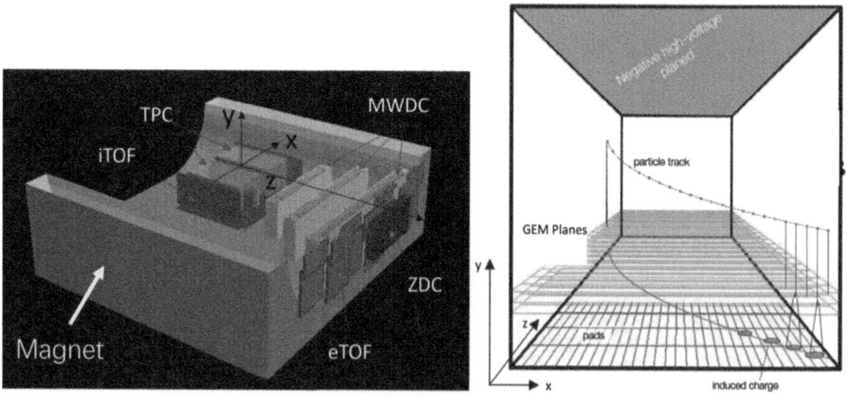

Fig. 5 (Left) The layout of the CEE experimental setup [3], (Right) Working principle of TPC [7]

points) for a step size of 1 mm. Each point actually contains also the information of true positions (x, y, z), momentum (p_x, p_y, p_z), energy deposited, track index, etc.

4.2 CEE TPC Digitization and Hit Reconstruction

Electron-ion pairs created at each step of energy loss are evaluated and uniformly distributed along the track direction. After that, we drift and diffuse electrons along longitudinal (time direction) and transverse direction (parametrizing diffusion with Gaussian functions) till the first GEM plane. Now each electron can make an avalanche (using gain histogram obtained from Garfield++ simulation) in the GEM layers. Assuming the gain as Gaussian, the charge is geometrically added to each pad (segmented anode into the rectangular sensors) as shown in the top panel of Fig. 6. After this, we treat these e-ion pairs to simulate the analog-to-digital conversion (ADC) using a single pad response function evaluated using SAMPA chip analysis [8]. Finally, we stored the fired pad index, ADC signal on each pad, discrete-time (timebin), etc.

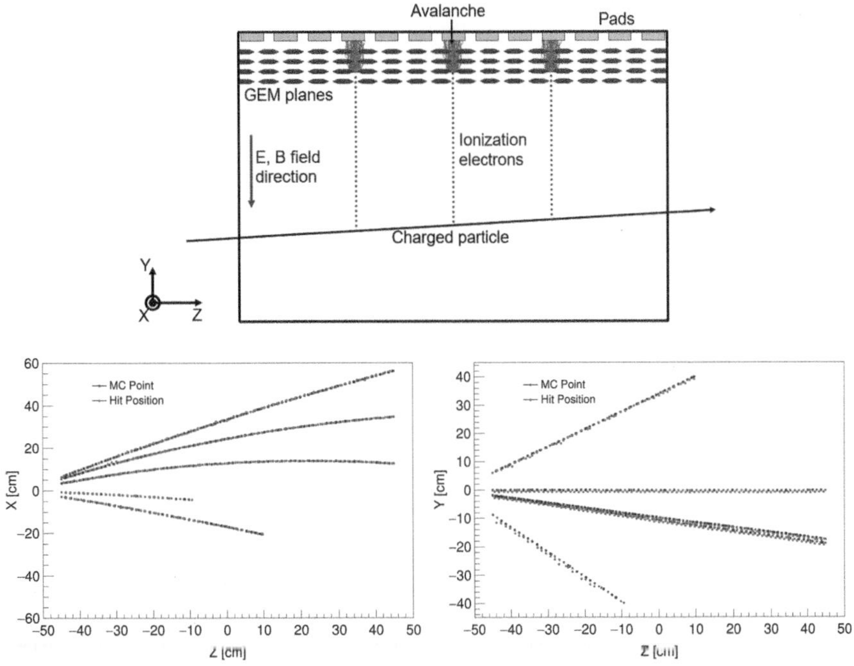

Fig. 6 (Top) Charged particle in the TPC, creating ionization, drift, diffusion, amplification in GEM, an induced charge on pads [3]. (Bottom left) MC X-Z points (Blue) and reconstructed XZ points (Magenta) (Bottom right) MC Y-Z points (Blue) and reconstructed YZ points (Magenta)

For the reconstruction of the hit position, we first evaluate the Z-position from the row index (rowId) corresponding to pad fired, then X-position is evaluated by using charge (on pads) center of gravity (COG) method, while Y-position is reconstructed using the charge in each timebin as described in the expressions below.

$$\mathrm{Z} = \mathrm{padheight} * (\mathrm{rowId\text{-}37}),\ \ \mathrm{X} = \frac{\sum \mathrm{x_i} * \mathrm{ADC_i}}{\sum \mathrm{ADC_i}},\ \ \mathrm{Y} = \frac{\sum \mathrm{timebin_i} * \mathrm{ADC_i}}{\sum \mathrm{ADC_i}} \tag{2}$$

The pad-plane has 200 pads of size 5 mm along X-direction and 75 pads of size 12 mm (height) along Z-direction. Therefore, we have 75 rows along the pad height direction (the 37th row corresponds to $Z_{rec} = 0$). The reconstructed X-Z and Y-Z hit position of the MC points are shown in the bottom left and bottom right panel of Fig. 6 respectively.

4.3 Spatial Resolutions and Design of a New Pad-Plane

Spatial resolutions is studied as a function of track angle with the z-direction (θ_x) and also with the drift direction (θ_y) as shown in the left, center, and right panels of Fig. 7. In the center of Fig. 7, the region where σ_x is increasing should not be considered because our main goal is to minimize the uncertainty in the coordinate measurement which ultimately will improve the tracking performance [9]. This can be done by minimizing the angle between tracks and pad direction. The new design shown in the top panel of Fig. 8, consists of four sectors with trapezoidal pads, and 45-degrees lines separating the consecutive sectors. In this configuration, tracks are always parallel to the pad direction, this omits the region of increasing σ_x with θ_x. The digitization is completed and pads fired correspond to MC points are shown in the bottom panel of Fig. 8. The spatial resolution studies are still ongoing for this new geometry.

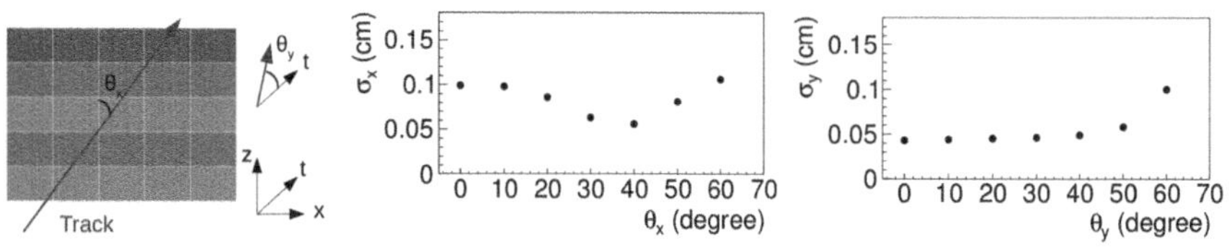

Fig. 7 (Left) Track making angle θ_x from z-direction and θ_y with the drift direction. (Center) At low θ_x, tracks create single pad cluster (σ_x constant), then as θ_x increases more pads fired improves σ_x, then at very θ_x cluster becomes very wider which further increases σ_x. (Right) σ_y stays constant with θ_y below 40 degree due to the clusters in fixed timebins and then increases due to wider clusters [3]

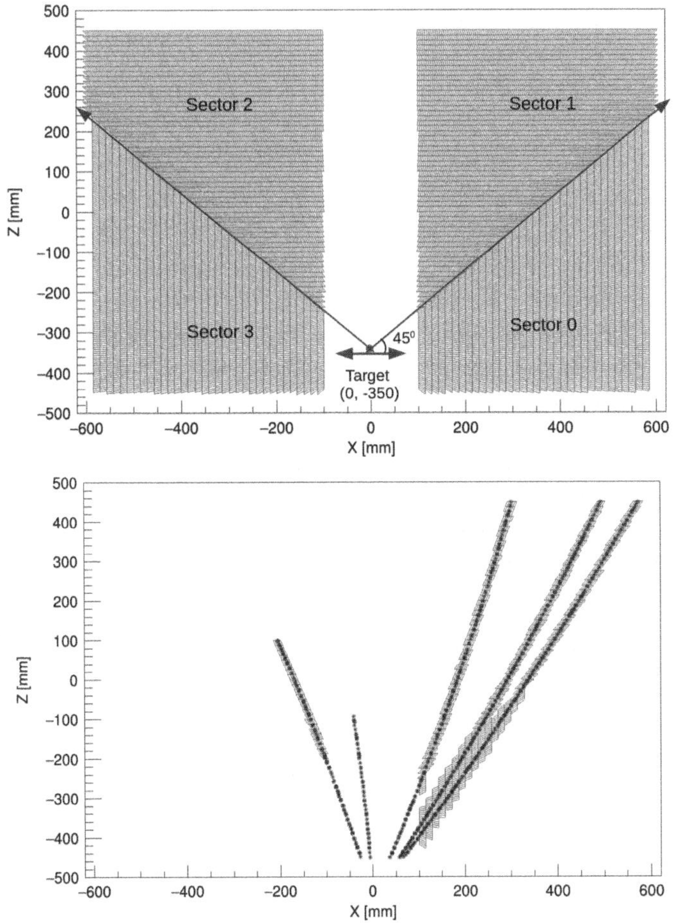

Fig. 8 (Top) New design of the pad plane consisting 4 sectors with trapezoidal pads in which tracks are parallel to pad direction. (Bottom) MC points created by the incident tracks (blue) and pads fired after the digitization

5 Tracking

Reconstruction of the trajectory of charged particles to the primary vertex from the hit positions is known as tracking. There are two main tasks: Track finding and Track fitting. Track finding (also known as pattern recognition) includes finding the hits that belong to track candidates. There are several algorithms e.g. Combinatorial Algorithm, Cellular Automaton, Hough Transform, etc. The CEE TPC is using Cellular Automaton (CA) for the track finding [10]. Track fitting includes fitting of tracks and estimation of track parameters using Kalman filter (genfit2 package) [11]. Quality of the track is based on the chi-square value of the Kalman fit. The

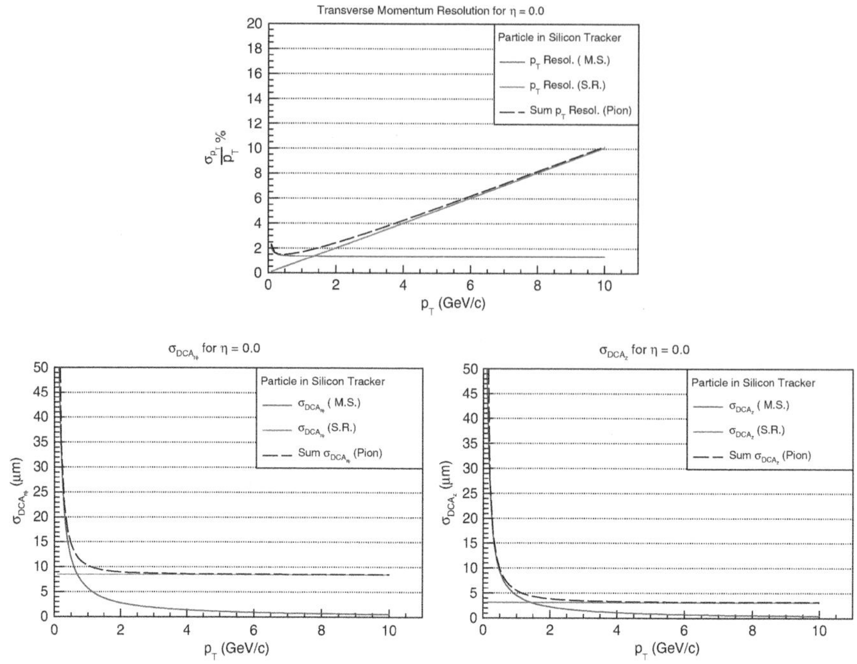

Fig. 9 (Top) p_T resolution as a function of p_T, (Bottom left) $DCA_{r\phi}$ as a function of p_T, and (Bottom right) DCA_Z as a function of p_T due to spatial resolution (S.R.) and multiple scattering (M.S.) for the geometry shown in Fig. 3

most important outcomes of tracking are p_T and Distance of Close Approach (DCA) resolutions. Figure 9 shows the resolutions (p_T, $DCA_{r\phi}$, DCA_z) estimated for the geometry in Fig. 3 using analytical formulas [9]. A good resolution is required to reconstruct the secondary particles, which decay close to the primary vertex e.g. D^0 meson ($c\tau = 123\,\mu m$) [12, 13]. Resolutions are optimized using MC simulations.

6 Analysis

After track reconstruction, particle identification (PID) can be performed with the detector setup using different methods namely energy loss versus momentum, time of flight, Cherenkov method [14]. Once we have the PID information several physics studies can be performed e.g. higher moment analysis of net-proton distributions for which identification of protons and antiprotons in each event is mandatory [15]. Similarly, the invariant mass analysis for the heavy-flavour candidates can be done using vertexing and PID information to reduce the combinatorial background [12]. The study can be performed on the simulated data to obtain the physics performance

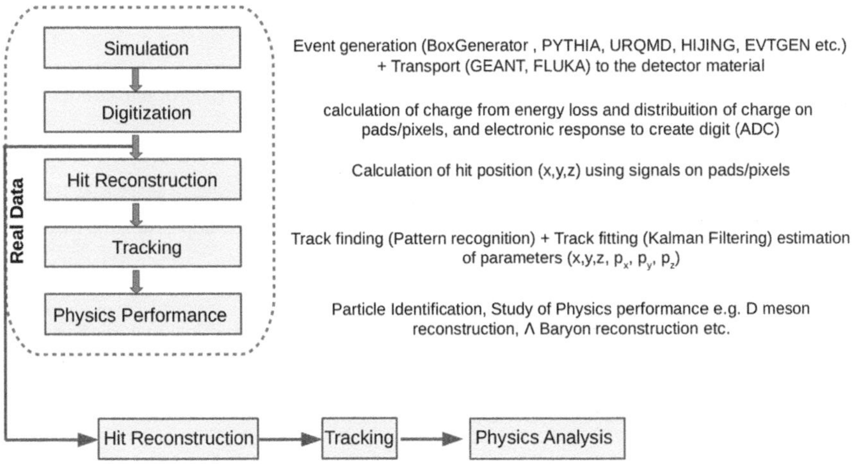

Fig. 10 Chain from detector simulation to data analysis in High Energy Physics Experiments

of the experimental setup while on the real data to extract the experimental physics results [16]. The whole chain of detector simulation is described in the diagram of Fig. 10.

7 Summary

The paper describes the chain of the detector simulation together with the important performance plots at each step using an example of the TPC for the CEE experiment. The design of TPC is yet to optimize based on the performance of spatial resolution using different readout pad sizes and shapes (rectangular, trapezoid, zig-zag, etc.). The basic concept of track finding and fitting are explained in this paper, however, further work is required to optimize the tracking performances for the experimental setup. The study reported here can be useful in designing the detector and physics performance study of the experiment.

Acknowledgements The work presented related to CEE TPC simulation is done at the IMP, CAS, China. I would like acknowledge Dr. Hao Qiu, Dr. Yapeng Zhang, Yi Zhu, Chenlu Hu, other members of the group to complete TPC simulation successfully. I would also like to acknowledge the ALICE Bari group for helping me to understand the tracking of charged particles.

References

1. M. Al-Turany, F. Uhlig, FairRoot framework. PoS **ACAT08**, 048 (2008). https://doi.org/10.22323/1.070.0048
2. S. Agostinelli et al., [GEANT4], GEANT4-a simulation toolkit. Nucl. Instrum. Meth. A **506**, 250–303 (2003). https://doi.org/10.1016/S0168-9002(03)01368-8
3. Z.H.U. Yi et al., Cluster reconstruction for the CSR External-target Experiment (CEE) time projection chamber. Nucl. Phys. Rev. **38**(4), 416–422 (2021). https://doi.org/10.11804/NuclPhysRev.38.2021021
4. P.A. Zyla et al., [Particle Data Group], Review of particle physics. PTEP **2020**(8), 083C01 (2020). https://doi.org/10.1093/ptep/ptaa104
5. M. Valentan et al.: Generalization of the Gluckstern formulas II: multiple scattering and non-zero dip angles. Nuclear Inst. & Methods Phys. Res. Sect. A-accelerators Spectrom. Detect. Ass. Equip. **606**, 728–742 (2009)
6. F. Ragusa, An introduction to charged particles tracking. Italo-Hellenic School of Physics (2006)
7. A. Delbart, et al., TPC Review and associated electronics. Workshop on A TPC for MeV Astrophysics Polytechnique, April 12th–14th (2017)
8. E. Jastrzembski, D. Abbott, J. Gu, V. Gyurjyan, G. Heyes, B. Moffit, E. Pooser, C. Timmer, A. Hellman, SAMPA based streaming readout data acquisition prototype. arXiv:2011.01345 [physics.ins-det]
9. Z. Drasal, W. Riegler, An extension of the Gluckstern formulae for multiple scattering: Analytic expressions for track parameter resolution using optimum weights. Nucl. Instrum. Meth. A **910**, 127–132 (2018). https://doi.org/10.1016/j.nima.2018.08.078. arXiv:1805.12014 [physics.ins-det]
10. R. Frühwirth Are Strandlie: Pattern Recognition. Tracking and Vertex Reconstruction in Particle Detectors (2021)
11. T. Bilka, N. Braun, T. Hauth, T. Kuhr, L. Lavezzi, F. Metzner, S. Paul, E. Prencipe, M. Prim, J. Rauch, et al., Implementation of GENFIT2 as an experiment independent track-fitting framework. arXiv:1902.04405 [physics.data-an]
12. S. Acharya, et al., [ALICE], Azimuthal correlations of prompt D mesons with charged particles in pp and p–Pb collisions at $\sqrt{s_{NN}} = 5.02$ TeV. Eur. Phys. J. C **80**(10), 979 (2020). https://doi.org/10.1140/epjc/s10052-020-8118-0. arXiv:1910.14403 [nucl-ex]
13. M. Faggin [ALICE], Vertexing detectors and vertexing performance in Run 2 in ALICE. PoS **LHCP2020**, 042 (2021). https://doi.org/10.22323/1.382.0042. arXiv:2008.13530 [physics.ins-det]
14. C. Lippmann, Particle identification. Nucl. Instrum. Meth. A **666**, 148–172 (2012). https://doi.org/10.1016/j.nima.2011.03.009. arXiv:1101.3276 [hep-ex]
15. M.M. Aggarwal, et al., [STAR], Higher moments of net-proton multiplicity distributions at RHIC. Phys. Rev. Lett. **105**, 022302 (2010). https://doi.org/10.1103/PhysRevLett.105.022302. arXiv:1004.4959 [nucl-ex]
16. [ALICE], ALICE upgrade physics performance studies for 2018 Report on HL/HE-LHC physics. ALICE-PUBLIC-2019-001

Contributed Papers

The Study of the Frequency of Bubble Oscillation in R-12 Superheated Emulsion for Neutron -Gamma Discrimination

Suraj Ali, Mala Das, and Pabitra Kumar Paul

1 Introduction

The superheated liquid maintains its liquid state at a temperature above its boiling point and it is a metastable state of the liquid. It moves to the stable (vapour) state after the formation of a critical size bubble by the energy deposition of energetic radiation or particle. The vapour bubble expands very rapidly and emits an acoustic pulse which can be detected by an active device like a microphone, piezoelectric transducer, etc. A superheated emulsion detector (SED) consists of a large number of superheated liquid drops of micron radius in a viscoelastic gel [1] or soft polymer [2]. This type of detector is useful in the detection of neutrons [3–6], gamma-rays [7, 8], and other charged particles [9, 10]. The customized SED is extensively used in the detection of the WIMP (Weakly Interacting Massive Particle), a favoured candidate of cold dark matter [11, 12]. The WIMP and neutron both initiate nucleation by the elastic collision with the nuclei of the detector material after satisfying the bubble nucleation conditions [13]. The threshold energy of such a detector depends on the operating temperature and pressure of the detector and the type of liquid used in the detector. Neutrons may initiate nucleation at high threshold energy or low operating temperature of the detector while the detector is insensitive to gamma-

S. Ali · M. Das (✉)
Saha Institute of Nuclear Physics, 1/AF Bidhannagar, Kolkata 700064, India
e-mail: mala.das@saha.ac.in

S. Ali
e-mail: suraj.ali@muralidhargirlscollege.ac.in

S. Ali · P. K. Paul
Department of Physics, Jadavpur University, 188 Raja S. C. Mallick Rd, Kolkata 700032, India
e-mail: pabitrak.pal@jadavpuruniversity.in

S. Ali
Muralidhar Girls' College, P411/14 Gariahat Road, Kolkata 700029, India

R. N. Patra (ed.), *Advanced Radiation Detector and Instrumentation in Nuclear and Particle Physics*, Springer Proceedings in Physics 282,
https://doi.org/10.1007/978-3-031-19268-5_4

rays. It becomes sensitive to gamma-rays above some specific operating temperature and it was found that the sensitivity starts at 41.9 °C [14], 45 °C [15], and 38.5 °C [16] for R-12 (CCl_2F_2; $b.p. = -29.8$ °C), R-114 ($C_2Cl_2F_4$; $b.p. = 3.7$ °C) and R-134a ($C_2H_2F_4$; $b.p. = -26.3$ °C) liquid SEDs respectively. The gamma-rays are the dominant background in the WIMPs detection experiments at the low threshold energy of the detector. It is therefore important to discriminate the gamma-ray to use this low threshold detector in the detection of neutron or WIMP in presence of gamma-ray background. Das et al. [17] have studied the neutrons and gamma-rays-induced pulses from the nucleation of R-114 SED and discriminated them using the maximum amplitude and power of the pulses. Mondal et al. [18] observed a significant difference between amplitude, power, and frequencies of the neutrons and gamma-rays-induced pulses using R-12 liquid in the audio frequency region (20 Hz to 20 kHz). Felizardo et al. [19] and Barnabe-Heider et al. [20] found the fundamental harmonic of neutron-induced events at a frequency between 0.45–0.75 kHz and 20–40 kHz using R-115 and R-610 liquid detectors. In the present work, we have tried to discriminate neutrons and gamma-rays-induced events from R-12 SED using Pvar and F.F. distributions in the high-frequency region. This high-frequency pulse is produced at the very early stage of the nucleation and carries the information about the value of Linear Energy Transfer (LET) of the nucleating particle [21].

2 Condition of Nucleation

The bubble nucleates after forming a bubble of critical radius (R_c) [22] represented by Eq. (1) and the embryo larger than or equal to R_c grows to visible vapour bubbles whereas the embryo smaller than R_c collapses back to the liquid state.

$$R_c = \frac{2\sigma(T)}{(P_v - P_l)} \tag{1}$$

To form a bubble of critical radius the energetic particle should deposit energy along the effective path length (L_{eff}) and the energy must be greater than or equal to the critical energy ($E_c(T)$). $E_c(T)$ can be expressed by Eq. (2) [23] and its variation at different temperatures is shown in Fig. 1,

$$E_c = 4\pi R_c^2\left(\sigma - T\frac{\partial\sigma}{\partial T}\right) + \frac{4\pi}{3}R_c^3\rho_v(h_v - h_l) - \frac{4\pi}{3}R_c^3(P_v - P_l) \tag{2}$$

where $\sigma(T)$ is the liquid-vapour interfacial tension at temperature T, P_v and P_l are the vapour and liquid pressure, h_v and h_l are the specific enthalpies of the vapour bubble and liquid respectively and ρ_v is the vapour density. The effective path length can be written as bR_c, where b is the nucleation parameter.

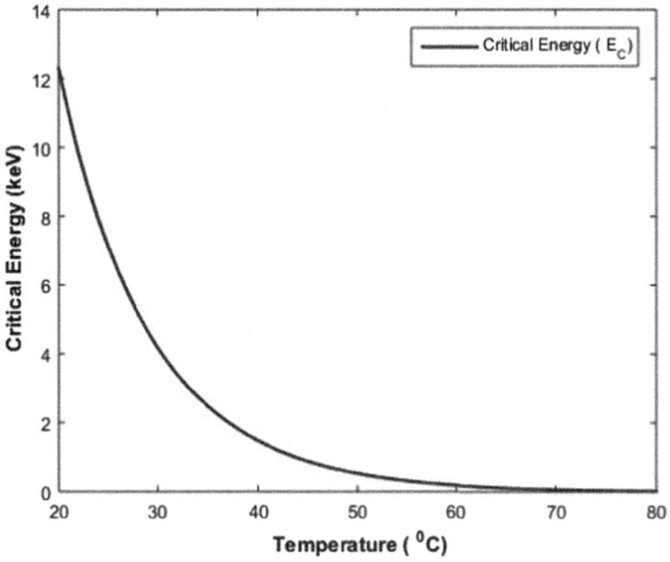

Fig. 1 The variation of critical energy for bubble formation as a function of temperature

3 Present Experiment and Analysis

The SED was fabricated using superheated R-12 liquid and aquasonic gel matrix. The experiments have been performed with the fabricated SED, one in the presence of a neutron source ($^{241}Am - Be$) of activity 10 mCi with average and maximum neutron energy of 4.2 MeV and 11 MeV respectively. Another in the presence of a gamma-ray source (^{137}Cs) of activity 5 mCi with gamma-ray energy of 0.662 MeV. The experiment with a neutron and gamma-ray source was performed at 35 °C and 55 °C respectively. The temperature of the experiment was controlled and measured using a temperature sensor and a controller (METRAVI, DTC 200) with precision ±1 °C. The nucleated acoustical pulses were collected using an AE sensor (AE - WSα by Physical Acoustics Corporation) placed on the top of the detector. The LabView hardware and software were used to store the pulses into data files. At the beginning of the analysis, the noises were rejected by visual selection [24] and only the traces decaying in nature were considered as actual bubble nucleated pulses. First, the pulses were analysed in the time domain and the Pvar variable was collected from each pulse. The Pvar variable is the summation of the square of the amplitude of the signal which is proportional to the energy deposited by the energetic particle. The fast Fourier transformation (FFT) was done for the collected pulses. In the FFT spectrum, the frequency with maximum power is defined as the fundamental frequency (F.F.). Discrimination can be quantified by the percentage of discrimination and it has been calculated by subtracting the overlapping area of F.F. distribution of neutron and gamma-rays-induced pulses, from the total area of each F.F. distribution.

4 Result and Discussions

The Pvar distributions in Fig. 2 show that the neutrons and gamma-rays-induced pulses are merged. The distribution of F.F. variable corresponding to neutrons and gamma-rays-induced signals is presented in Fig. 3. The distribution shows that the F.F. of the pulses from gamma-rays is lower than those of the neutrons and allows discrimination of 83.47% of neutrons from gamma-rays. A possible explanation is that Pvar has been calculated by summing the square of the voltages over the total time span of the signal so here both the amplitude and time information is present but the F.F. only contains the frequency or time information of the pulses. The neutron-induced pulses have low time span and high amplitude and gamma-rays-induced pulses have high time span and low amplitude as a result both the pulses have the same Pvar value. The F.F. value discriminates between the neutrons and gamma-rays-induced pulses as the recoil nucleus from neutrons deposits the whole energy within the critical radius due to its higher LET and lower ranges and hence produces the high-frequency pulses. On the other hand, the gamma-rays-induced pulses are predicted to grow at a slower rate due to the energy deposition over an extended track of the electrons and produce low-frequency pulses.

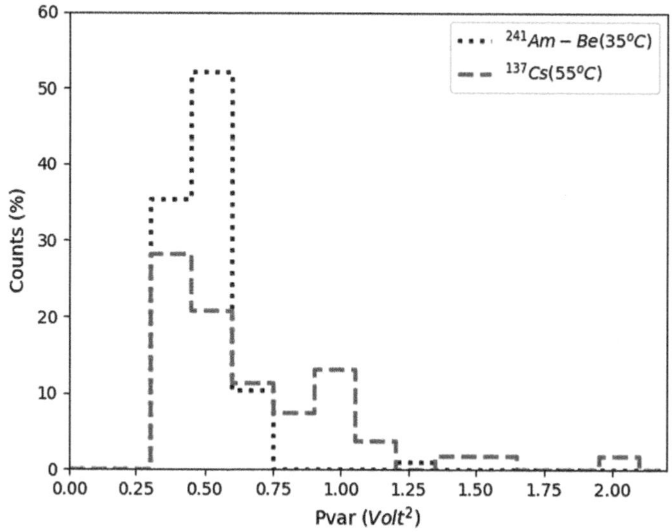

Fig. 2 The distribution of Pvar of the neutrons and gamma-rays-induced acoustical pulses

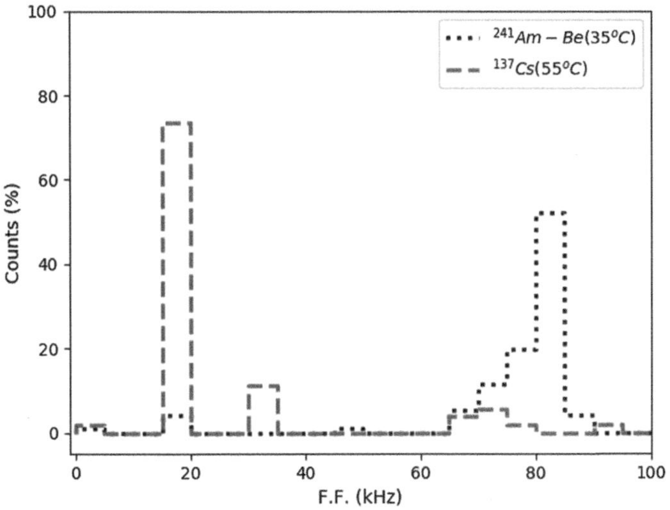

Fig. 3 The distribution of F.F. of the neutrons and gamma-rays-induced acoustical pulses

5 Conclusion

In this work, we have studied the frequency spectrum of the bubble nucleated pulses from neutrons and gamma-rays in R-12 SED. Although Pvar parameters of both types of signals are almost the same, the F.F. parameter significantly discriminates the neutrons-induced pluses from the gamma-rays-induced pulses.

Acknowledgements The authors would like to thank Mr. Nilanjan Biswas, Dr. Sunita Sahoo, SINP, and Mr. Sudipta Barman, Mechanical Workshop, SINP for the assistance during the experiment. The authors are thankful to Prof. Maitreyee Nandy, SINP and Health Physics Unit, Variable Energy Cyclotron Centre for providing $^{241}Am - Be$ source. The authors are grateful to CAPP - II project, SINP-DAE for the financial support.

References

1. R.E. Apfel, The superheated drop detector. Nucl. Instrum. Methods **162**, 603–608 (1979)
2. H. Ing, H.C. Birnboim, A bubble-damage polymer detector for neutrons. Nucl. Tracks Radiat. Meas. **8**, 285–288 (1984)
3. R.E. Apfel, B.T. Chu, J. Mengel, Superheated drop nucleation for neutron detection. Appl. Sci. Res. **38**, 117–122 (1982)
4. B. Mukherjee, D. Rybka, D. Makowski, T. Lipka, S. Simrock, Radiation measurement in the environment of FLASH using passive dosimeters. Meas. Sci. Technol. **18**, 2387 (2007)
5. P.K. Mondal, B.K. Chatterjee, Characteristics of acoustic emissions during nucleation of superheated droplets. Phys. Lett. A **375**, 237–244 (2011)
6. F. d'Errico, Radiation dosimetry and spectrometry with superheated emulsions. Nucl. Instrum. Methods Phys. Res. B **184**, 229–254 (2001)

7. F. d'Errico, R. Nath, R. Nolte, A model for photon detection and dosimetry with superheated emulsions. Med. Phys. **27**, 401–409 (200)
8. B. Roy, M. Das, S.C. Roy, B.K. Chatterjee, Threshold temperature for γ-ray detection in superheated drop detector. Radiat. Phys. Chem. **61**, 509–510 (2001)
9. S.L. Guo et al., High energy heavy ion tracks in bubble detectors. Radiat. Meas. **31**, 167–172 (1999)
10. S. Tsuda, T. Sawamura, A. Homma, M. Narita, Pressure dependence of neutron detection sensitivity in a superheated drop detector. Radiat. Prot. Dosim. **87**, 87–92 (2000)
11. E. Behnke et al., Improved limits on spin-dependent WIMP-proton interactions from a two liter CF_3I bubble chamber. Phys. Rev. Lett. **106**, 021303 (2011)
12. M. Felizardo et al., First results of the phase II SIMPLE dark matter search. Phys. Rev. Lett. **105**, 211301 (2010)
13. C. Amole et al., Dark matter search results from the PICO-60 C_3F_8 bubble chamber. Phys. Rev. Lett. **118**, 251301 (2017)
14. R. Sarkar, P.K. Mondal, B.K. Chatterjee, The gamma ray detection threshold temperature of different superheated droplet detectors. Appl. Radiat. Isot. **139**, 127–130 (2018)
15. M. Das, B. Roy, B.K. Chatterjee, S.C. Roy, Use of basic principle of nucleation in determining temperature-threshold neutron energy relationship in superheated emulsions. Radiat. Phys. Chem. **66**, 323–328 (2003)
16. S. Sahoo, S. Seth, M. Das, The threshold of gamma-ray induced bubble nucleation in superheated emulsion. Nucl. Instrum. Methods Phys. Res. A **931**, 44–51 (2019)
17. M. Das, S. Seth, S. Saha, S. Bhattacharya, P. Bhattacharjee, Neutron-gamma discrimination by pulse analysis with superheated drop detector. Nucl. Instrum. Methods Phys. Res. A **622**, 196–199 (2010)
18. P.K. Mondal, S. Seth, M. Das, P. Bhattacharjee, Study of low frequency acoustic signals from superheated droplet detector. Nucl. Instrum. Methods Phys. Res. A **729**, 182–187 (2013)
19. M. Felizardo et al., The SIMPLE phase II dark matter search. Phys. Rev. D **89**, 072013 (2014)
20. M. Barnabé-Heider et al., Response of superheated droplet detectors of the PICASSO dark matter search experiment. Nucl. Instrum. Methods Phys. Res. A **555**, 184–204 (2005)
21. S. Ali, M. Das, Discrimination of neutron and gamma ray induced nucleation events at high frequency in R134a superheated emulsion. Nucl. Instrum. Methods Phys. Res. A **1025**, 166186 (2022)
22. F. Seitz, On the theory of the bubble chamber. Phys. Fluids **1**, 1–13 (1958)
23. A.G. Tenner, Nucleation in bubble chambers. Nucl. Instrum. Methods **22**, 1–42 (1963)
24. S. Seth, M. Das, Radiation linear energy transfer and drop size dependence of the low frequency signal from tiny superheated droplets. Nucl. Instrum. Methods Phys. Res. A **837**, 92–98 (2016)

Simulations of Multi-layer GEM Systems from Single to Quadruple GEMs

Aera Jung, Yong Ban, Dayong Wang, Yue Wang, and Licheng Zhang

1 Introduction

In 1997 Fabio Sauli introduced Gas Electron Multiplier (GEM) technology [1], a new generation of multi-wire proportional chambers allowing for the detection and localization of ionizing radiation with sub-millimeter accuracy and high event rate capability. Although it was originally invented as a preamplifier to help the then novel microstrip gas chambers cope with the high particle rate, development was subsequently pursued by several groups, and GEMs were successfully used in many experiments. Finally GEMs became a basic feature of Gas Pixel Detectors [2].

With a multi-GEM layer structure a very high effective gain can be attained with each GEM layer working at an individually much lower gain thus requiring a lower voltage and avoiding discharge problems. This is the major advantage of the GEM technology. Hence, in the present paper, we present comparative simulation results for single, double, and triple layer GEM systems, along with some preliminary triple and quadruple layer results. This paper also aims to provide guidelines for the

A. Jung (✉) · Y. Ban · D. Wang · Y. Wang · L. Zhang
School of Physics, Peking University, 209 Chengfu Road, Haidian District, Beijing, China
e-mail: aera@pku.edu.cn

Y. Ban
e-mail: bany@pku.edu.cn

D. Wang
e-mail: dayong.wang@pku.edu.cn

Y. Wang
e-mail: 1801210129@pku.edu.cn

L. Zhang
e-mail: zhanglic@pku.edu.cn

A. Jung
University of Hawaii, Honolulu, USA

R. N. Patra (ed.), *Advanced Radiation Detector and Instrumentation in Nuclear and Particle Physics*, Springer Proceedings in Physics 282,
https://doi.org/10.1007/978-3-031-19268-5_5

multi-GEM layer's tendencies based on the author's simulations. Due to limited space it does not cover the underlying physical or scientific theories.

2 Detector Configuration in the Simulation

This simulation is based on the Garfield++ software toolkit [3]. Unfortunately, Garfield++ cannot compute complex electric fields, therefore first the commercial software ANSYS [4] is separately applied to the GEM detector geometry to create a model of the detector's electric field using the finite element method (FEM). FEM divides an object of interest into smaller elements and then calculates the electric potential and field vectors at the nodes [5]. To simplify this process, ANSYS only simulates a limited region (2 GEM holes) which is then symmetrically extrapolated into an infinite sheet so as to negate edge effects. The ANSYS outputs are imported into Garfield++ which then performs simulations of the electrons as they move though the detector. For our GEM geometry, we considered a standard hexagonal GEM foil with a pitch of 140 μm, a outer hole diameter of 70 μm, an inner hole diameter of 50 μm, and a Kapton thickness of 50 μm with 5 μm copper conductive layers on both sides, and filled with a gas mixture of 70% argon and 30% carbon dioxide.

Table 1 shows our detector configuration in the simulation. Our simulated detector configuration was derived by referring to the detector designs of Bachmann [6] and Rajendra Nath Patra's [7] papers.

Bachmann's paper [6]: For the single, double, and triple GEM simulations, voltages across each GEM were kept identical, with transfer and induction fields of 3.5 kV/cm and drift fields of 2 kV/cm. The drift distance is kept at a consistent 3 mm, however transfer (spacing between the different GEM foils) and induction distances (the

Table 1 Summary of detector configuration in the simulation

	Drift distance [mm]	Transfer distance [mm]	Induction distance [mm]	HV divider		
				Drift field [kV/cm]	Transfer field [kV/cm]	Induction field [kV/cm]
Bachmann's paper [6]						
Single	3		1/2	2		3. 5
Double	3	1/2	1/2	2	3. 5	3. 5
Triple	3	1/2	1/2	2	3. 5	3. 5
Rajendra Nath Patra's paper [7]						
Triple	4.8	2	2	Originally the top panel of Fig. 3 in his paper		
Quadruple	4.8	2	2	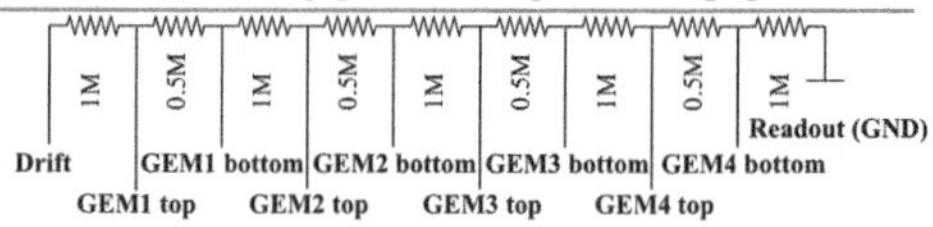		

spacing between last GEM foil and the readout) of both 1 and 2 mm were used. For example, our double detector's gaps are 3 mm from drift to first GEM, 1 or 2 mm between the first and second GEM, and 1 or 2 mm from the second GEM to the readout plate.

Rajendra Nath Patra's paper [7]: On triple and quadruple GEM simulations, the drift gap is 4.8 mm and the detector's transfer and induction gaps are all 2 mm. That is, our detector's gaps are 4.8 mm from drift to first GEM, 2 mm between the first and second GEM, another 2 mm between the second and third GEM, and finally 2 mm from the third GEM to the readout plate. We also used a high voltage (HV) divider as shown in Table 1. The HV divider circuit distributes the voltage across the detector so as to generate the needed electric field geometry. Specifically, it produces outputs that supply the drift, the top and bottom conductive layer of each GEM, and the readout plate which is set as the electrical ground of the HV system.

3 Results

We simulated the GEM detector using the parameters mentioned above and focused on determining the gain, electron transparency, efficiency, spatial resolution, and energy resolution by varying the electron flow into single layer, double layer, and triple layer GEM systems based on S. Bachmann's paper. This has been described in Sect. 3.1. We also present preliminary results of the gain, electron transparency, efficiency, and spatial resolution for triple and quadruple layer GEMs based on Rajendra Nath Patra's paper in Sect. 3.2.

For these simulations, the penning transfer ratio is manually set to 0.57 as given in [8] and the gas is set to 1 atm pressure at room temperature. For initiating the avalanche process, a single electron with 0.5 eV energy is placed 2.995 mm above the top of the first GEM.

3.1 Single, Double, and Triple GEMs

Detailed studies of gain, spacial resolution, energy resolution, electron transparency and efficiency for single, double, and triple systems have been published in a 37th ICRC (International Cosmic Ray Conference) 2021 proceeding paper [9]. We reproduce here some data relevant and comparative to the present study.

Figure 1 shows the simulation results of single, double, and triple GEMs for gain, spatial resolution, energy resolution, electron transparency, and efficiency.

The gain is given by the number of electrons created by each primary electron that reaches the anode. The gain varies with almost every parameter involved in the operation of a GEM. In this simulation, these include GEM voltage and the number of GEM layers (i.e. single, double, or triple GEMs).

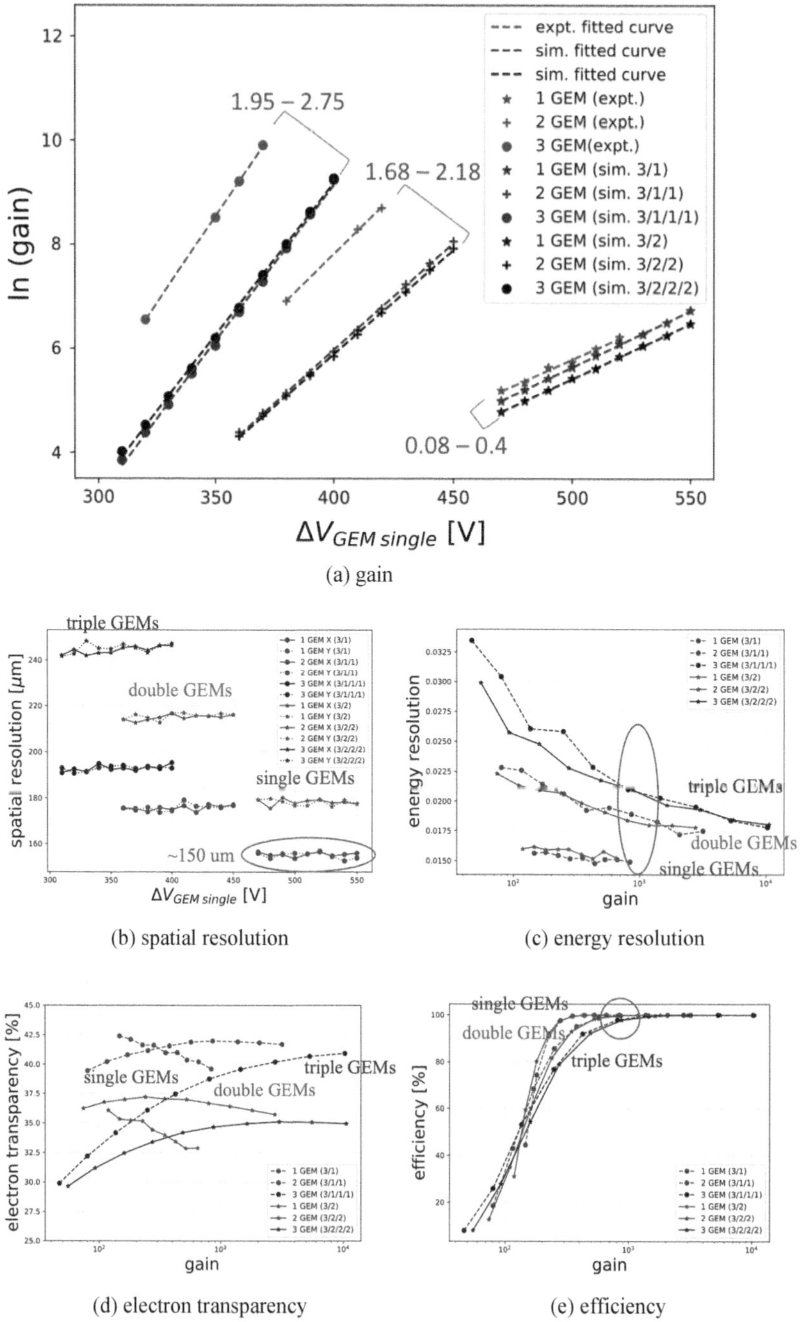

Fig. 1 Simulation results of single, double and triple GEMs

Green is the experimental data originally from Bachmann's paper by eye, blue is the simulation model with a transfer and induction gap of 1 mm, and black is the simulation model with a transfer and induction gap of 2 mm. Moreover, the star symbol is a single GEM, the plus symbol is a double GEM, and the circle symbol is a triple GEM. There is only a small difference between the experimental gain value and the simulation gain value for a single GEM. The difference between experimental and simulation gain values of double GEM is larger than that of single GEM and the difference is larger still for a triple GEM. Therefore, as the number of layers of the GEM increases, the difference between experimental and simulation results increases. This is a very well known issue in which the gain seen in experiments exceeds that which can the obtained in simulations results [10, 11]. Many groups are doing research on this topic, but we still do not understand exactly why this discrepancy exists. So far, simulations cannot achieve the nano, let alone sub-nano, scale resolution required to accurately reflect the short time scale in which the microscopic interaction phenomena occur. This is assumed to be the reason why the gain seen in simulations is lower.

The spatial resolution is one of the key parameters for tracking systems and can be extracted from the width of the residual distribution reached on the anode plate.

For a single GEM with a transfer and induction gap of 1 mm (red circle), the spatial resolution is about 150 μm. It can be seen that the distance from the first GEM to the readout plate increases by about 15 μm/mm. I.e. a triple GEM with a 2 mm transfer and induction gap has a spatial resolution of 240 μm ($= 150 + 15 \times 6$).

The energy resolution is central for GEM detectors working in proportional mode and other devices aiming for a measurement of the deposited energy. From the mean and the sigma of the gain distribution, the energy resolution has been estimated, as shown in the middle row right side panel of Fig. 1. As the layers of the GEM increase, the energy resolution deteriorates near a gain of 1000 as marked in magenta.

The electron transparency is the ratio of secondary electrons arriving at the readout to that of all secondary particles (especially electrons). Therefore, when we run the simulation, we know all the secondary particles and we know how many secondary particles (especially electrons) arrive at the readout. Of course removing the electrons from atoms generates positive ions however as these are not of interest we will not discuss their behavior.

Overall the total electron transparency of single, double, and triple GEMs is between 29 and 42.5% depending on the chosen gain and detector configuration.

The efficiency describes the probability of a trespassing particle to yield the expected signal and, if applicable, to overcome a threshold value needed have this signal recognized. This detector efficiency was simulated to estimate the performances of the VFAT3 transfer function of the induced anode current given by Garfield++ [12]. We used the same threshold for single, double, and triple GEMs. This threshold value was set slightly higher than the noise.

The efficiency increases with gain and then comes to a plateau after a certain gain value is reached. The smaller number of GEM layers are faster, arriving soonest at 100%. It means a single GEM has a gain of around 500, double a gain of around 900, and triple a gain of around 1100.

3.2 Triple and Quadruple GEMs

Figures 2 and 3 show the simulation results of triple and quadruple GEMs for gain, spatial resolution, total electron transparency, and efficiency.

Gain

In Fig. 2 left panel, green is the experimental data originally from Rajendra Nath Patra's paper [7] by eye and blue is our simulation models. Moreover, the star symbol is for a triple GEM system while the plus symbol is for a quadruple GEM. The difference between simulation and experimental data for a triple GEM detector is about 0.88–1.56 and quadruple GEM is about 2.65–3.44. Seen in the above results for single, double, and triple GEMs, as the number of GEM layers increases, the difference between experimental and simulation results increases. However, the difference between the experimental and simulation values of the triple GEM in Fig. 1a is larger than that of the triple GEM in Fig. 2 left panel. When the gain is about 10,000, the delta GEM voltage (difference between the top and bottom GEM plates) is the same as 400 V, but the electric fields are different. That is, the drift field of the triple GEM in Fig. 1a is 2 kV/cm, and the transfer field and induction field are 3.5 kV/cm. The drift field of the triple GEM in Fig. 2 left panel is 1.5 kV/cm, and the transfer field and induction field are 3.7 kV/cm.

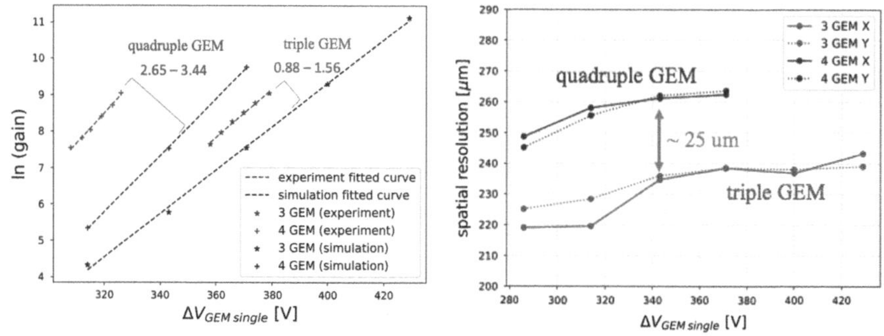

Fig. 2 Left: Gain as a function of GEM voltage between top and bottom GEM electrodes for triple and quadruple GEMs. Green is the experimental data from Rajendra Nath Patra's paper [7] by eye and blue is the simulation model with triple (star symbol) and quadruple (plus symbol) GEMs. Right: Spatial resolution as a function of GEM voltage between top and bottom GEM electrodes for triple (red) and quadruple (blue) GEMs

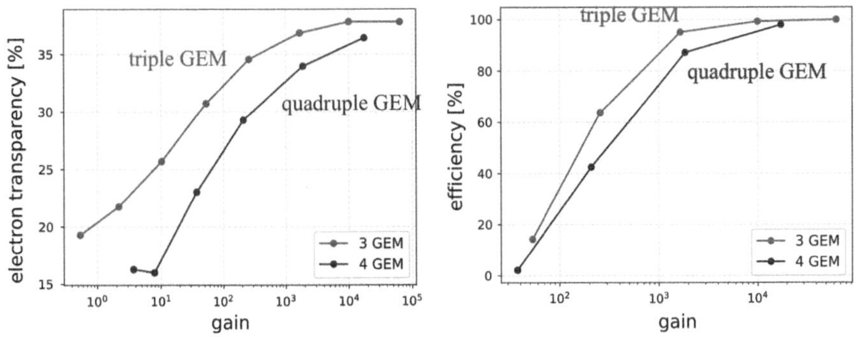

Fig. 3 Left: Total electron transparency as a function of gain for triple (red) and quadruple (blue) GEMs. Right: Efficiency as a function of gain for triple (red) and quadruple (blue) GEMs

Spatial resolution

In Fig. 2 right panel, red is triple GEM and blue is quadruple GEM. The difference in spatial resolution between triple and quadruple GEMs is about 25 μm. For single, double, and triple GEM results, distance from the first GEM to the readout plate increases by about 15 μm/mm. Therefore, the expected difference is around 30 μm. Hence, our simulated value of 25 μm is close.

Electron transparency

For gain values between ~100 and 7000, the electron transparency is about 27–38%. In addition, the triple GEM's electron transparency is a little higher than the quadruple GEM's electron transparency. This is similar to the previous single, double, and triple GEMs results.

Efficiency

The efficiency increases with gain and then comes to a plateau after a certain gain value is reached, as shown in the right panel of Fig. 3. The triple GEM layer is faster, arriving soonest at 100%.

It is not recommended to directly compare the simulation efficiency with the experimental efficiency value. Because the electronics are different. However, the efficiency of the experimental value also reaches 100% in the triple GEM first.

4 Summary

We have performed a comparative simulation study of single, double, triple and quadruple GEM systems with Garfield++ and ANSYS. For single, double, and triple GEMs, as the number of GEM layers is increased, the gain difference between experimental results and simulation increases. On the other hand, energy resolution deteriorates as the number of GEM layers is increased while maintaining the system at a constant gain. The spatial resolution becomes poorer as the distance between the first GEM and the anode increases. However, this difference is only about 15 μm/mm. While there are some differences in electron transparency, single, double, and triple GEMs are pretty much the same. Lastly, the fewer number of GEM layers the faster the system reaches 100% efficacy.

For triple and quadruple GEMs, we simulated the gain, spatial resolution, electron transparency, and efficiency. These results show similar tendencies to those seen in single, double, and triple GEMs. Based on these studies, we plan to conduct simulations of various delta GEM voltages and also increase the number of events so as to study the energy resolution and related properties in comparison with triple and quadruple detectors. Lastly, our group has also already submitted our quadruple GEM's performance results to the Radiation Detection Technology and Methods Journal [13].

In general increasing the gain also increases the noise. Therefore, from all the simulated results, the spatial resolution, energy resolution, and efficiency become poorer as the number of stages is increased. Nevertheless, we still prefer to use multi-layer GEMs because the deteriorating values are not particularly large.

Acknowledgements The authors have been extremely fortunate to have the support, mostly via email and other discussions, and encouragement of many individuals as well as various organizations. We are thank to Dr. Jeremie Alexandre Merlin, Dr. Rajendra Nath Patra, Dr. Rob Veenhof, Dr. Supratik Mukhopadhyay, Dr. Lia Lavezzi, Dr. Heinrich Schindler, Luis Felipe Ramirez Garcia, and the CERN service-desk. We would like to convey our gratitude to each and every one of them.

References

1. F. Sauli, Nucl. Instr. Meth. **A386**, 531 (1997)
2. S.D. Pinto, J. Spanggaard, Comments: Presented at IBIC 2013, Oxford. arXiv:1309.2908
3. Garfield++ CERN. https://garfieldpp.web.cern.ch/garfieldpp/
4. ANSYS. https://www.ansys.com/
5. John W. Robinson et al., Nucl. Instr. Meth. A **810**, 37–43 (2016)
6. S. Bachmann et al., Nucl. Instr. Meth. A **479**, 294–308 (2002)
7. R.N. Patra et al., Nucl. Instr. Meth. A **906**, 37–42 (2018)
8. O. Sahin et al., Nucl. Instr. Meth. A **768**, 104–111 (2014)
9. A. JUNG et al., PoS, ICRC2021 186 (2021)
10. S. Swain et al., Springer Singapore (2018). https://doi.org/10.1007/978-981-10-7665-7_24
11. P.K. Rout et al., JINST **16**(02), P02018 (2021)
12. T. Maerschalk, Ph.D. thesis (2016). https://cds.cern.ch/record/2291028
13. L. ZHANG et al., Simulation study of the performance of quadruple-GEM detector. Radiat. Detect. Technol. Methods (submitted)

A Simulation of Primary Ionization for Different Gas Mixtures

R. Kanishka, Supratik Mukhopadhyay, Nayana Majumdar, and Sandip Sarkar

1 Introduction

The gaseous detectors [1] have been used in the particle physics experiments such as CMS [2, 3], INO [4, 5], LHCb [6, 7] and in the upgraded ALICE-TPC [8, 9]. The work presented describes the study of primary ionization using different radiation sources in the argon based gas mixtures. The primary ionization initiates transportation and amplification of ions and electrons in the detector that has been used to find the discharge probability in GEM-based detectors [10]. This motivated us to study primary ionization using different radiation sources to find their responses in the various gas mixtures. The simulation of primary ionization in argon based gas mixtures to obtain energy and spatial information has been presented here. Four type of gas mixtures have been used. Ar-CO_2 (90-10), Ar-CO_2 (70-30) [11, 12] and Ar-CO_2-CF_4 (45-15-40) have been used as these gases have good transport properties whereas addition of CF_4 shows fast detection [13]. Ar-CO_2 (80-20) was studied additionally. The numerical model, results and conclusions are presented in next sections.

R. Kanishka (✉) · S. Mukhopadhyay · N. Majumdar · S. Sarkar
Applied Nuclear Physics Division, Saha Institute of Nuclear Physics, 1/AF, Bidhannagar, Kolkata 700064, India
e-mail: kanishka.rawat.phy@gmail.com

S. Mukhopadhyay
e-mail: supratik.mukhopadhyay@saha.ac.in

N. Majumdar
e-mail: nayana.majumdar@saha.ac.in

S. Sarkar
e-mail: sandip.sarkar@saha.ac.in

Homi Bhabha National Institute, Training School Complex, Anushaktinagar, Mumbai 400094, India

R. N. Patra (ed.), *Advanced Radiation Detector and Instrumentation in Nuclear and Particle Physics*, Springer Proceedings in Physics 282,
https://doi.org/10.1007/978-3-031-19268-5_6

2 Numerical Model

Geant4 [14] and heed++ [15, 16] toolkits have been used to simulate the primary ionization in argon based gas mixtures with the radiation sources such as alpha, muon and ^{55}Fe. Geant4 toolkit is capable of simulating the passage of particles through the matter. Thus, it is feasible to simulate the radiation source, matter for interaction and the detectors of an experiment in geant4. A particle interacts, decays and get scattered after each step because of the different physics processes used in the simulation. Different geant4 packages have been used for low energy and radioactive physics. For alpha simulation EMPenelope, EMLivermore, Photo Absorption Ionization (PAI), PAI-Photon have been used. EMPenelope and EMLivermore are electromagnetic physics lists that have been used for primary ionization of low energy (below 1 GeV) physics. PAI, PAI-Photon physics lists are capable for transport of fast charged particles, describe number and positions of primary ionization and for low energy primary particles. In ^{55}Fe simulation excitation of atoms produces characteristic x-rays and auger electrons therefore G4AtomicDeexcitation physics list has been used. The processes like fluorescence, particle-induced x-ray emission, auger electron emission and auger cascade are included in this physics list. A simulation of the ionization pattern produced by the track of the particles using heed++ toolkit in the gaseous detectors has been done in this paper.

3 Results

We describe the different simulations in the next sub-sections.

3.1 Alpha Simulation

Four different gas mixtures such as Ar-CO_2 (90-10), Ar-CO_2 (80-20), Ar-CO_2 (70-30), Ar-CO_2-CF_4 (45-15-40) have been utilized for the simulation. Argon acts as a target for ionization, whereas CO_2 and CF_4 are photon and electron quenchers respectively. Argon atoms have higher ionization energy than CO_2 and de-excitation of argon atoms ionize CO_2 via photo absorption. An alpha beam (^{241}Am) of 5.6 MeV energy was shot from the center as shown in the Fig. 1. The dimensions of the gas volume were 5 cm along x and y, and 20 cm along z direction. EMPenelope, EMLivermore, PAI, PAI-Photon were the physics lists used for the simulation. 10000 events were shot in the argon based gas mixtures. Figure 2 shows the x, y, z co-ordinates of the primaries obtained when alpha of 5.6 MeV energy was simulated in the four gas mixtures. Ar-CO_2 (90-10) produces more counts than other gas mixtures since the argon ratio is highest than in the other gas mixtures. Figure 2 also shows that x and y co-ordinates observe Gaussian distribution since the number of events are

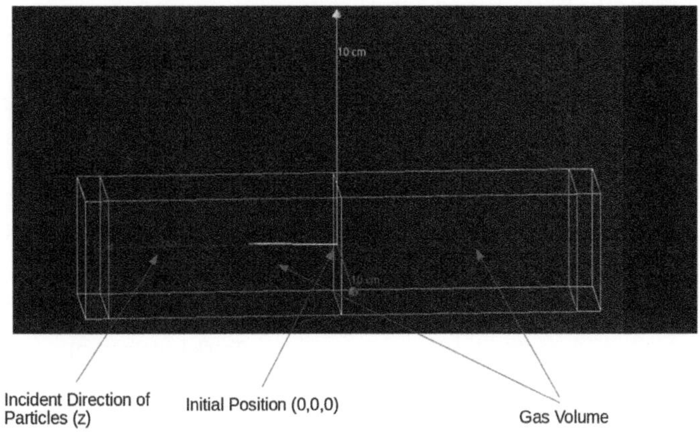

Fig. 1 The geant4 display of alpha of 5.6 MeV energy that was simulated in Ar-CO_2 (70-30). Thick dotted line shows geant4 hits

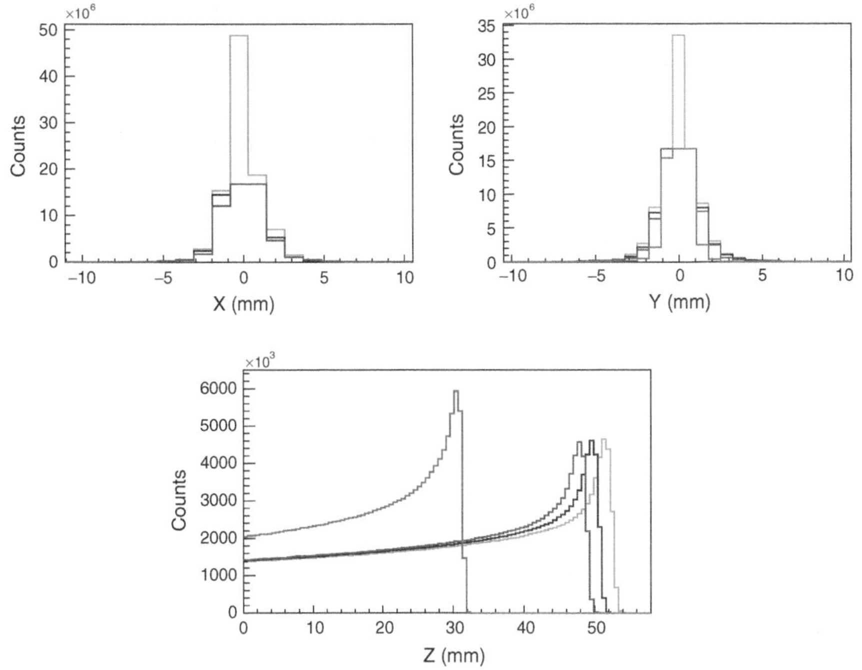

Fig. 2 The x, y and z co-ordinates of the primaries obtained from alpha of 5.6 MeV energy in four different argon based gas mixtures, Ar-CO_2 (90-10), Ar-CO_2 (80-20), Ar-CO_2 (70-30), Ar-CO_2-CF_4 (45-15-40) respectively

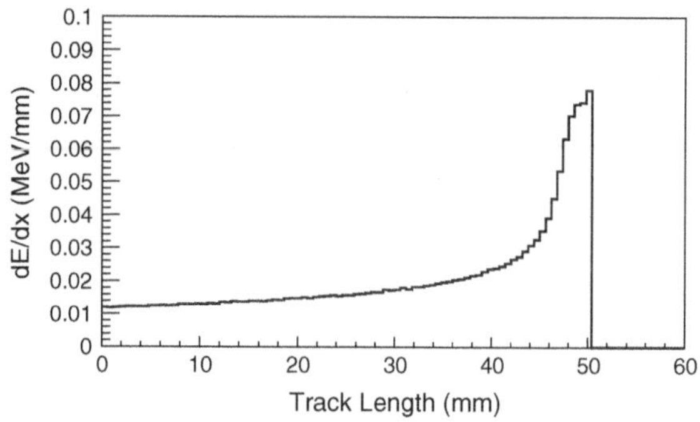

Fig. 3 The Bragg curve obtained from alpha particle simulation in Ar-CO_2 (70-30)

large. The x and y co-ordinates are centered at zero as the particles were shot at x = y = 0. In Fig. 2, the z co-ordinate indicates the range. The particles create primaries while travelling along z direction. When the argon ratio in the gas mixture was increased the number of primaries also increased. The smallest range has been observed for Ar-CO_2-CF_4 (45-15-40) gas mixture. The range of alpha particle in these four gas mixtures lie between 3.3 and 5.4 cm [17]. Figure 3 shows the Bragg curve for Ar-CO_2 (70-30) gas mixture. A constant deposition of energy of the particle has been observed across the track length and was maximum at the end of the trajectory, because the energy varies $1/v^2$ where v is the velocity of the particle. Thus as particle slows down it deposits most of its energy.

3.2 Muon Simulation

For simulation of 1 GeV muons the same dimensions of the gas volume, physics lists as that of alpha particle have been chosen. 10000 events were shot from x = 0, y = 0, z = −10 cm in the Ar-CO_2 (70-30) gas mixture. Figure 4 shows the position co-ordinates obtained when muon of 1 GeV energy placed at (0, 0, −10 cm) was simulated in gas mixture. The figure shows that x and y co-ordinates observe a Gaussian distribution. Due to the large range of muons the z co-ordinate is uniformly distributed. The number of primaries/cm obtained were 64.85.

3.3 ^{55}Fe Simulation

A radioactive source ^{55}Fe of 0.00589 MeV energy placed at (0, 0, 0 cm) was simulated in Ar-CO_2 (70-30) gas mixture. The number of events and dimensions of the gas volume were the same as used in alpha source. The physics list chosen was G4AtomicDeexcitation. ^{55}Fe captures the electron (let's assume) from K-shell, then electron from L-shell fills that vacancy thereby producing ^{55}Mn–$K_{\alpha 1}$ (x-rays).

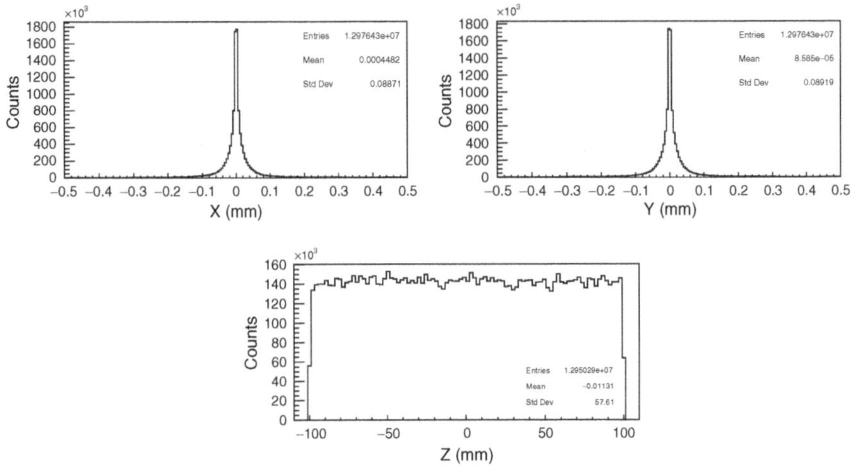

Fig. 4 The x, y, z co-ordinates of the primaries obtained from muon of 1 GeV energy placed at (0, 0, –10 cm) in Ar-CO_2 (70-30) gas mixture

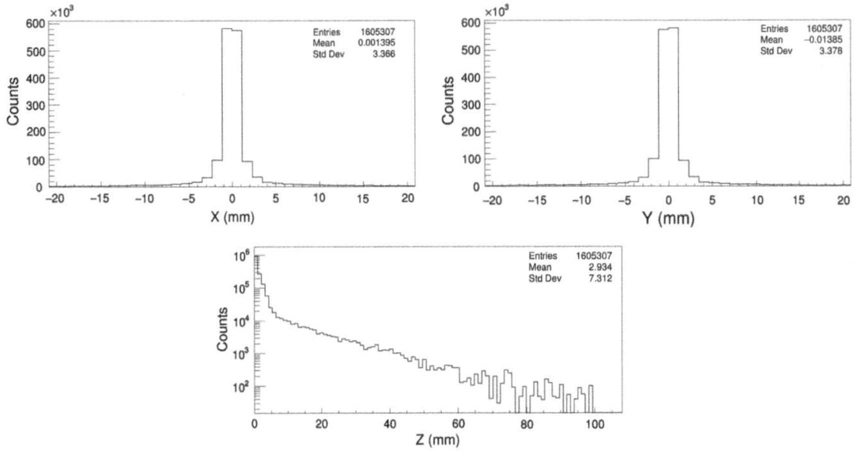

Fig. 5 The x, y, z co-ordinates of the primaries obtained from 0.00589 MeV ^{55}Fe in Ar-CO_2 (70-30) gas mixture

If electron from M-shell fills the vacancy of K-shell then ^{55}Mn–K$_{\beta 1}$ are released. Sometimes the emitted photons eject an outer electron and is known as auger electron emission. A neutrino is also emitted during electron capture process. Figure 5 shows the x, y, z co ordinates of the primaries obtained from ^{55}Fe simulation in Ar-CO_2 (70-30) gas mixture. Due to the loss of energy, the counts in the z co-ordinate decreases when primaries travel along its track.

10000 events of gamma were also shot in Ar-CO_2 (70-30) gas mixture using heed++. Left of Fig. 6 shows energy peaks i.e., K$_{\alpha 1}$ and K$_{\beta 1}$ of x-rays (0.00589 and

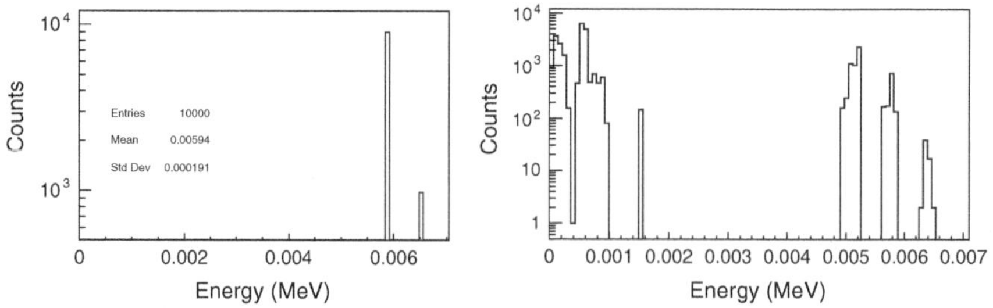

Fig. 6 Left: $K_{\alpha 1}$ and $K_{\beta 1}$ x-rays (0.00589 and 0.00649 MeV) obtained from gamma source in Ar-CO_2 (70-30) gas mixture using heed++. Right: Last two peaks refers to $K_{\alpha 1}$ and $K_{\beta 1}$ x-rays obtained from ^{55}Fe simulation in Ar-CO_2 (70-30) gas mixture using geant4. Note: On right plot other peaks refers to energy peaks of electron emission

0.00649 MeV) obtained when gamma was simulated in Ar-CO_2 gas mixture. These energy peaks exactly match with the standard results [18]. The last two peaks on right plot shows energy peaks of ^{55}Mn ($K_{\alpha 1}$ and $K_{\beta 1}$) using geant4 from ^{55}Fe simulation in Ar-CO_2 (70-30) gas mixture. The other peaks on right plot refers to energy peaks of electron emission. This is due to the inclusion of G4AtomicDeexcitation physics list in geant4 toolkit that includes fluorescence, particle-induced x-ray emission, auger electron and auger cascade emission. Since the physics list was used in geant4 therefore the heights of the two x-ray peaks in right plot obtained from geant4 differ from the left plot (heed++ toolkit).

4 Conclusions

We simulated alpha, muon and ^{55}Fe sources in different argon based gas mixtures to obtain their response since the properties of these sources are different. The geant4 and heed++ toolkits have been used for the studies to obtain spatial and energy information. Bragg curve was obtained for alpha particle simulation. In another analysis, ^{55}Fe captures the electron to emit x-rays while the electron re-arrangement takes place. The energy peaks of x-rays obtained closely match with the standard results.

Acknowledgements Authors acknowledge the organizers of Rapid2021 and University of Jammu. R. Kanishka thank SINP for the computing service. We also thank P. K. Rout and other colleagues of ANP division.

References

1. F. Michael et al., Gas ionization detectors, in *Handbook of Radioactivity Analysis*, 4th edn. (Academic Press, 2020), pp. 245–305
2. D. Abbaneo et al., Characterization of GEM detectors for application in the CMS muon detection system, in *2010 IEEE Nuclear Science Symposium, Medical Imaging Conference, and 17th Room Temperature Semiconductor Detectors Workshop, RD51-NOTE-2010-005* (2010), pp. 1416–1422. https://doi.org/10.1109/NSSMIC.2010.5874006
3. A. Colaleo, A. Safonov, A. Sharma, M. Tytgat, CMS technical design report for the muon endcap GEM upgrade, CERN-LHCC-2015-012, CMS-TDR013 (2015)
4. B. Satyanarayana, Design and characterisation studies of resistive plate chambers, Ph.D. thesis, Department of Physics, IIT Bombay, PHY-PHD-10-701 (2009)
5. A. Kumar et al., Invited review: physics potential of the ICAL detector at the India-based Neutrino Observatory (INO). Pramana—J. Phys. **88**, 79 (2017). https://doi.org/10.1007/s12043-017-1373-4
6. G. Bencivenni et al., A triple GEM detector with pad readout for high rate charged particle triggering. NIMA **488**, 493 (2002)
7. M. Alfonsi et al., The LHCb triple-GEM detector for the inner region of the first station of the muon system: construction and module-0 performance, in *IEEE Symposium Conference Record Nuclear Science*, vol. 1 (2004), pp. 5–9. https://doi.org/10.1109/NSSMIC.2004.1462057
8. B. Ketzer, A time projection chamber for high-rate experiments: towards an upgrade of the ALICE TPC. NIMA **732**, 237 (2013)
9. J. Adolfsson et al., The upgrade of the ALICE TPC with GEMs and continuous readout. JINST **16**, P03022 (2021)
10. P.K. Rout, R. Kanishka et al., Numerical estimation of discharge probability in GEM-based detectors. JINST **16**, P09001 (2021)
11. S. Bachmann et al., Discharge studies and prevention in the gas electron multiplier (GEM). NIMA **479**, 294 (2002)
12. P. Gasik et al., Charge density as a driving factor of discharge formation in GEM-based detectors. NIMA **870**, 116 (2017)
13. Badriya Al-Rashdi et al., Numerical study of transport properties of Ar-based ternary gas mixtures and their gain in GEM detectors. Appl. Math. Sci. **15**, 225–237 (2021)
14. S. Agostinelli et al., Geant4—a simulation toolkit. NIMA **506**, 250 (2003)
15. I.B. Smirnov, Detailed simulation of the initial ionization in gases. CERN Comput. Newsl. **226**, 13 (1996)
16. I.B. Smirnov, Modeling of ionization produced by fast charged particles in gases. NIM A **554**, 474–493 (2005)
17. A.K. Solomon, *Experimental Nucleonics*, ed. by E. Bleuler, G.J. Goldsmith, vol. 117 (New York, Rinehart, 1953), pp. 87–88
18. http://nucleardata.nuclear.lu.se/toi/nuclide.asp?iza=260055

Simulating Response of a Liquid Scintillation Detector to Gamma and Neutrons

S. Das, V. K. S. Kashyap, and B. Mohanty

1 Introduction

In rare event search experiments, fast neutrons are one of the most dominant contributors to the background. They can be emitted from the surrounding rock materials via (α,n) processes, spontaneous fission of heavy element impurities (U and Th), and through cosmic ray interactions. EJ-301 is a popular liquid scintillation detector that detects fast neutrons. Gamma and neutrons interact with the scintillator via electron recoil and nuclear recoil [1]. Hence we have to calibrate both the energies separately. It is difficult to get mono-energetic neutron sources. Unfolding can be a way of calibrating the neutron energy even if the spectrum is continuous. To do this, we need the response of the detector to the neutron source and a detector response matrix that connects the incident energy of neutrons to the light output of the scintillator. The light output equation for incident neutrons with a specific energy distribution can be written using the response matrix as, $N = R \times \phi$ [2]. Where ϕ represents the energy spectrum of the radioactive source, N represents measured detector response in optical counts, and R represents the response matrix. In unfolding method, we aim to find ϕ using detector response N and response matrix R. A Geant4 simulated liquid scintillation detector and its response to gamma and neutrons have been presented in this work, along with the reconstruction of true Am-Be spectra using unfolding method.

S. Das (✉) · V. K. S. Kashyap · B. Mohanty
School of Physical Sciences, National Institute of Science Education and Research, Jatni 752050, India

Homi Bhabha National Institute, Training School Complex, Anushaktinagar, Mumbai 400094, India
e-mail: sudipta.das@niser.ac.in

R. N. Patra (ed.), *Advanced Radiation Detector and Instrumentation in Nuclear and Particle Physics*, Springer Proceedings in Physics 282,
https://doi.org/10.1007/978-3-031-19268-5_7

Fig. 1 (left) EJ-301 detector coupled to a photo-multiplier tube. (right) Complete experimental setup consists of detector, power supply, digitizer, shielding, radioactive sources and PC

2 Experimental Setup

In the experiment, we have used a 2" $\times$ 2" EJ-301 liquid scintillation detector [3]. The detector shows excellent pulse shape discrimination properties between gamma and neutrons events. We have achieved a Figure Of Merit (FOM) value of ≈ 0.9 with an energy threshold at 90 keV_{ee} using charge integration method. The detector is coupled to an R7724 Photo Multiplier Tube (PMT) from Hamamatsu. The anode output is connected to a 14-bit CAEN V1730 digitizer which performs Digital Pulse Processing (DPP). The output of the digitizer is connected to the PC. The experimental setup is shown in the Fig. 1. The standard gamma and neutron sources used in our experiment are ^{22}Na, ^{137}Cs, ^{60}Co and ^{241}Am-^{9}Be respectively.

3 Detector Simulation in Geant4

We have simulated the same EJ-301 detector that we used in our experiment. The detector geometry has been constructed in Geant4 simulation. The dimensions of the detector are provided by the manufacturer [3]. 97% volume of the detector is filled with scintillator (hydrocarbon) and 3% inert gas (Argon is used in our simulation). The inside of the scintillator is coated with TiO_2, which has a reflectivity of 90% to ensure the light pulse reaches the PMT.

We place the detector above a granite base to mimic the experimental setup [4]. The radioactive sources are kept 10 cm away from the detector and are assumed to emit radiation isotropically. We have used the physics list—`G4OpticalPhysics` to simulate the production of optical photons. The scintillator properties like the value of the refractive index, absorption length, scintillation yield, scintillation time components, transition energies are taken from the EJ-301 datasheet [3]. We have also added the boundary properties for different interfaces (metal-dielectric and dielectric-

dielectric) inside the detector. To simulate the gamma and neutron interaction with scintillator the physics lists: `ElectromagneticPhysics` and `G4HPNeutron` [5], available in Geant4 has been used.

4 Determining the Resolution Parameter in Simulation

The resolution parameter of a scintillator measures the spread of generated optical photon numbers relative to the scintillation-yield [5] and can be written as, $\texttt{Spread=Resolution parameter} \times \sqrt{\texttt{Scintillation yield}}$. EJ-301 scintillator, shows a scintillation yield of 12000 optical photons per 1 MeV_{ee} energy deposition. In the simulation, we specify the resolution parameter of the scintillator by comparing the detectors' simulated and experimental gamma responses. The simulated gamma response for ^{22}Na spectrum is shown in the upper left panel of Fig. 2. We can see that it follows the Compton spectra with Compton edges broadened due to the resolution effect. Now we need to convert the X-axis in terms of detected optical photons to ADC channels. This is done to compare the simulated spectrum with experimental measurement. In the experiment, we have obtained detector responses to ^{137}Cs, ^{22}Na and ^{60}Co gamma sources. We have set the charge sensitivity of the digitizer to be 5 fC/LSB, and it has a precision of 14 bit (16383 channels). If N number of photons fall on the photocathode we can write,

$$\text{ADC channel} = \frac{\text{N} \times 0.26 \times 4.44 \times 10^4 \times 1.602 \times 10^{-19}}{5 \times 10^{-15}} = \text{N} \times 0.37, \tag{1}$$

where, the quantum efficiency of the photo-cathode is 26% and dynodes together create 4.44×10^4 secondary electrons on an average for each photo-electron. Using Eq. 1, we can relate the total number of optical photons falling on the photocathode to the ADC channels. ^{22}Na gamma response with different resolutions has been simulated and compared with the experimental measurement. A resolution parameter value of 5 in our simulation gives the best matching result. In Fig. 2 comparison between simulated and experimental gamma responses have been shown for ^{137}Cs, ^{22}Na and ^{60}Co spectra. The values of ratio plots close to one suggest that simulated spectra are consistent with the experimental one.

5 Simulating Neutron Response of the Detector and Reconstructing the True Am-Be Spectra

Neutrons interact with the scintillator via nuclear recoil. They can scatter with the detector nuclei (mostly proton in case of organic scintillator) both elastically and inelastically. Then the recoiled proton excites the electrons of scintillator molecules. When the electrons de-excite, it emits optical photons. `G4HPNeutron` has been

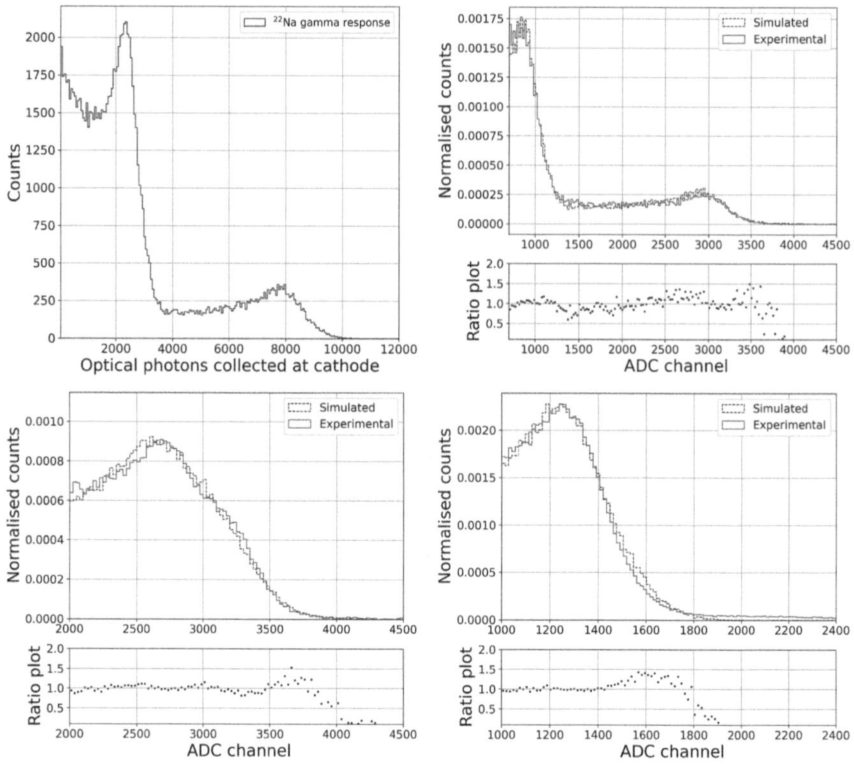

Fig. 2 (upper left) Simulated gamma response of detector to ^{22}Na gamma source in terms of optical photons collected at photo-cathode. (Upper right) Detector Response to ^{22}Na gamma source. (Lower left) Detector response to ^{60}Co gamma source. (lower right) Detector response to ^{137}Cs gamma source

used to simulate the detector response to neutrons. The simulated neutron response matrix of EJ-301 is shown in the upper left panel of Fig. 3.

The ISO spectrum of neutrons emitted from Am-Be sources follows the distribution shown in the upper right panel of Fig. 3. We generated this spectrum in Geant4 and made it fall on the scintillation detector. The neutrons scattered from the granite base can also reach the detector and act as an additional source. We can see that this effect modifies the spectra but very slightly.

The simulated response of the detector to Am-Be neutrons is shown in the lower panel of Fig. 3. Now using the response matrix, we have to unfold the simulated neutron response of the detector to obtain the input Am-Be spectrum (ISO Am-Be spectrum). Two unfolding methods have been used: (a) Gravel iterative method [2] and (b) RooUnfold method [6]. The reconstructed spectra obtained from the unfolding are shown in the left panel of Fig. 4 with the variation of $\chi^2/n.d.f$ value with iteration number in the right panel of the same figure.

We can see that both unfolding methods are consistent with the ISO spectrum.

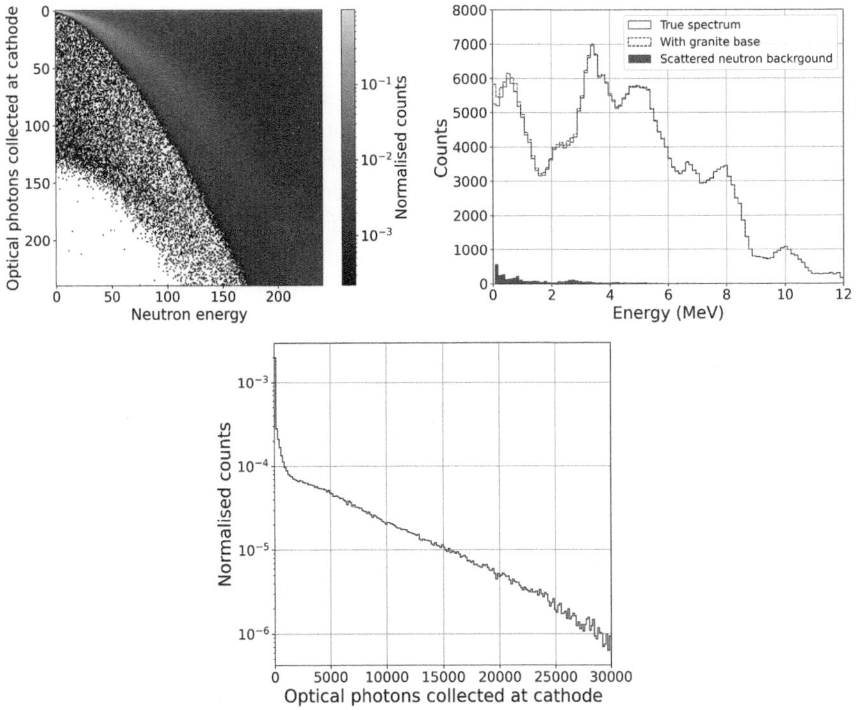

Fig. 3 (Upper left) Response matrix 240 × 240. (Upper right) Am-Be ISO neutron spectrum. (Lower middle) Simulated detector response to Am-Be neutron source in terms of optical counts

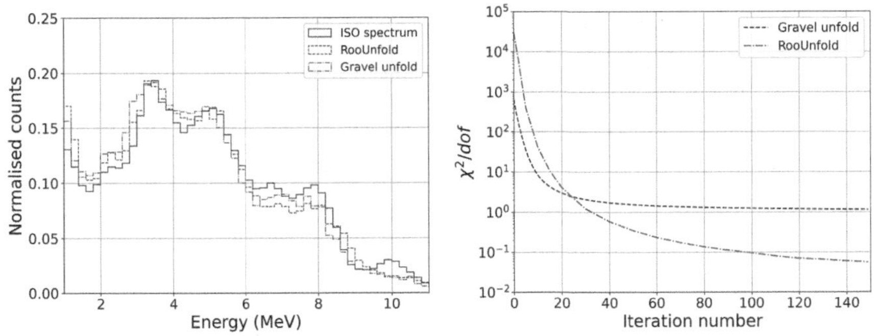

Fig. 4 (left) Unfolded Am-Be spectra from simulated detector response. (right) Variation of Chi square per degree of freedom with iteration number

6 Summary

The response of a 2" × 2" liquid scintillation detector to gamma and neutron has been simulated and neutron response to the Am-Be source has been unfolded. We see a good agreement between experimental and simulated gamma responses. We also see the reconstructed spectra obtained from the two unfolding methods are consistent to the ISO spectrum of the Am-Be source. The convergence of $\chi^2/n.d.f$ is much faster in Roo-Unfolding. It is also time-efficient. Our next plan is to experiment with the Am-Be source and unfold the experimental neutron response. Once done, the detector and its simulated neutron response matrix can be used to understand the fast neutron background from the measurement made at rare event experimental sites.

Acknowledgements The work has been carried out in the NISER neutron detector laboratory which has been set up by the support of DAE-India through the project Research in Basic Sciences—Dark Matter and SERB DST-India through the J. C. Bose Fellowship.

References

1. G.F. Knoll, Radiation detection and measurement, 4th ed. (Wiley, 2010)
2. Y. Chen, X. Chen, J. Lei, L. An, X. Zhang, J. Shao, P. Zheng, X. Wang, Unfolding the fast neutron spectra of a BC501A liquid scintillation detector using gravel method. Sci. China Phys., Mech. Astron. **57**(10), 1885–1890 (2014)
3. Eljen Technology, EJ-301, EJ-309—Neutron/Gamma PSD liquid scintillators—Eljen Technology (2021). https://eljentechnology.com/products/liquid-scintillators/ej-301-ej-309
4. S. Das, V. Kashyap, B. Mohanty, Energy calibration of EJ-301 scintillation detector using unfolding methods for fast neutron measurement. Nucl. Instr. Methods Phys. Res. Sect. A: Accel. Spectrometers Detect. Assoc. Equip. **1042**, 167405 (2022). https://doi.org/10.1016/j.nima.2022.167405. https://www.sciencedirect.com/science/article/pii/S016890022200701X
5. Geant4 Collaboration, Book for application developers (2019). https://indico.cern.ch/event/679723/contributions/2792554/attachments/1559217/2453758/BookForApplicationDevelopers.pdf
6. ATLAS Collaboration, Introduction to unfolding with rooUnfold—statistical methods documentation (2019). https://statisticalmethods.web.cern.ch/StatisticalMethods/unfolding/RooUnfold_01-Methods/

Numerical Evaluation of Resistive Plate Chamber

Subhendu Das, Jaydeep Datta, Nayana Majumdar, and Supratik Mukhopadhyay

1 Introduction

Resistive Plate Chamber (RPC) is a gaseous detector with parallel plate configuration, known for its excellent time and spatial resolution. The easy and inexpensive fabrication of large area coverage has made this detector suitable for many modern experiments. Its working principle relies upon the amplification of primary electrons, created in its gas volume due to passage of charged particles, under the influence of an externally applied electric field. It is generated by supplying high voltages to the conductive coating applied on the resistive plates. The performance of the RPC is largely governed by the field configuration where the electrical properties of the device components play an important role. In this work, we have numerically simulated the electric field distribution and dark current of RPC due to the applied voltage across its resistive electrodes to get an optimal performance of RPC.

2 Configuration and Working Principle

RPC is constructed with two parallel plates of high volume resistivity, such as glass or bakelite [1, 2]. The gap between the plates is sealed with spacers from all sides to make it a chamber for holding the active gas volume. A suitable gas mixture is circulated through the volume using nozzles fixed on the side-spacers. To maintain the uniformity of the gap between the plates, several button-shaped spacers are used within the gas volume. The outer surface of the resistive plates is coated with a thin layer of conductive paint where the high voltage supply is connected to create a

S. Das (✉) · J. Datta · N. Majumdar · S. Mukhopadhyay
Saha Institute of Nuclear Physics, A CI of Homi Bhabha National Institute, AF block, Sector I, Salt Lake, Bidhannagar, Kolkata 700064, West Bengal, India
e-mail: subhendudas456038@gmail.com

R. N. Patra (ed.), *Advanced Radiation Detector and Instrumentation in Nuclear and Particle Physics*, Springer Proceedings in Physics 282,
https://doi.org/10.1007/978-3-031-19268-5_8

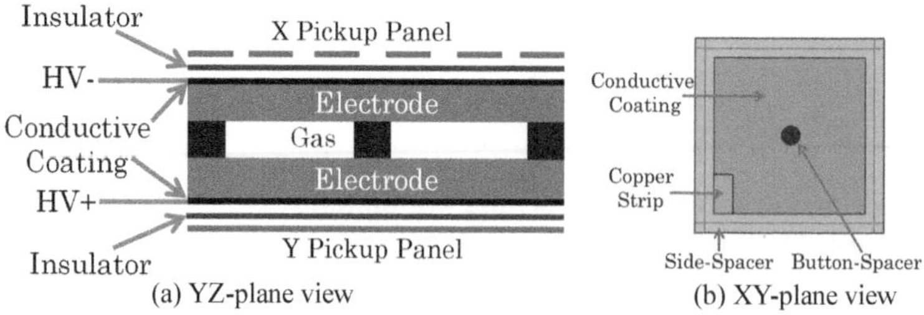

Fig. 1 Resistive Plate Chamber (RPC cross-section in (**a**) YZ-plane, (**b**) XY-plane)

uniform electric field across the gas-gap. This whole structure is sandwiched between two pickup panels for collecting the current signal of RPC which are insulated from the conductive coating by an insulator. The pickup panels are made by pasting copper strips on a firm structure of insulating materials. Two schematic planar views of RPC are shown in Fig. 1.

The primary electron-ion pairs created from the ionization of the gaseous molecules due to their interaction with the passing particle, drift towards respective electrodes under the action of the applied electric field. The primary electrons undergo multiplication through further interaction while moving towards the anode. The ions on the other hand drift slowly towards cathode. This movement of electrons and ions induces a current on the pickup panels [3]. The growth of the charges causes a drop in the voltage applied at the electrodes. The high volume resistivity of the electrodes ensures that such a drop is localized near the avalanche site. The ultra-violet absorbing component of the gas mixture absorbs the photons emitted by the excited gas molecules and prevents generation of secondary avalanches far away from the original one. This helps to keep the secondary avalanches located near the original one and in turn improves spatial resolution of the detector.

3 Simulation Model

The simulation of electric field configuration and dark current for the given design of RPC has been carried out using COMSOL Multiphysics [4]. A 3D-model of RPC with 3 mm thick resistive electrodes and a 2 mm gas-gap has been built for simulation. The side-spacers of width 10 mm and thickness 2 mm and a button-spacer of diameter 10 mm and thickness 2 mm have been considered. To avoid the possibility of discahrges across the edges of the electrodes we have considered the conductive coating has a dimension such that a gap of 1 cm is available at all the sides. Following the construction, a small copper strip for provision of high voltage supply has been considered at one corner of the conductive coating as shown in Fig. 1(b). The insulating layer and the pickup panel beyond the electrode have been considered

Table 1 The electrical properties of the materials

Material	Volume resistivity (Ω cm)	Relative permittivity
Gas	10^{18}	1
Glass electrode	10^{12}	10
Bakelite electrode	10^{10}	4
Polycarbonate spacer	10^{14}	3

in the present study, although there should not be any influence of these components on the field configuration.

The electric field and dark current in the RPC have been calculated by using the "electric current" module of the COMSOL, where the following equations have been solved.

$$\nabla \cdot \mathbf{J} = Q_{j,v} \qquad \mathbf{J} = \sigma \mathbf{E} + \mathbf{J_e} \qquad E = -\nabla V$$

where V and $\mathbf{E}$ are the voltage and electric field respectively, $\mathbf{J}$ is the current density, $\mathbf{J_e}$ is the external current density due to the electron avalanche, σ is the electrical conductivity of the material and $Q_{j,v}$ is the volumetric source of current. The electrical properties, volume resistivity and relative permittivity of the electrode and spacer materials considered for this numerical study have been given in Table 1.

4 Simulation Results

Here, all the results of the voltage distribution on the electrodes, dark current density in different components and electric field configuration within the active volume obtained for various electrical properties of the materials will be discussed.

4.1 Voltage Distribution

The voltage distribution on the resistive electrode depends on surface resistivity of the conductive coat along with the electrical properties of the electrode and the spacers. In order to study the effect of non-uniform surface resistivity resulting from the manual application of the conductive coat on the voltage distribution, it has been measured for a glass RPC of dimension 30 cm × 30 cm (coated area 28 cm × 28 cm) with 2 mm thick electrode and 2 mm gas-gap and the data have been used as input in the numerical model to simulate the potential distribution for the same RPC. In Fig. 2(a) and (b), the surface resistivity maps of the top and bottom glass electrodes have been displayed while in Fig. 3(a), the simulated voltage distribution of the top

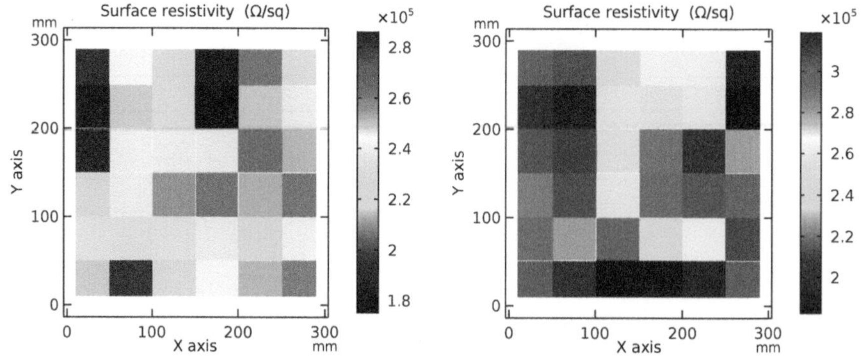

(a) (b)

Fig. 2 (**a**) Measured surface resistivity of top surface, (**b**) Measured surface resistivity of bottom surface

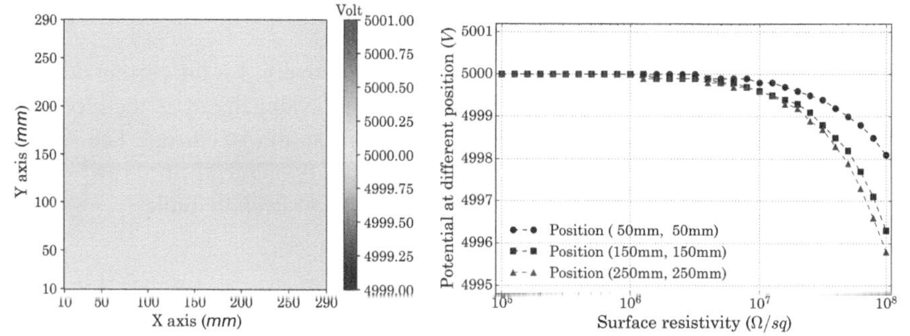

(a) (b)

Fig. 3 (**a**) Potential as a function of surface resistivity of conductive coating, (**b**) Simulated voltage distribution on top surface of the glass RPC

electrode has been shown. The result proves that the potential distribution remains uniform for the given non-uniformity in surface resistivity. Hence, in case of the present simulation study of the bakelite RPC, the effect of the surface resistivity on the voltage distribution has been assumed to be the same. It has been found that the potential at different position of the electrode can be held uniform for a wide range of variations in the surface resistivity (100 kΩ/□ – 1 MΩ/□) as shown in Fig. 3(b). It can be mentioned that the highly fluctuating surface resistivity may create non-uniform voltage distribution in one hand and shielding effect on induction of the output signal on the pickup panel lying outside [5] on the other.

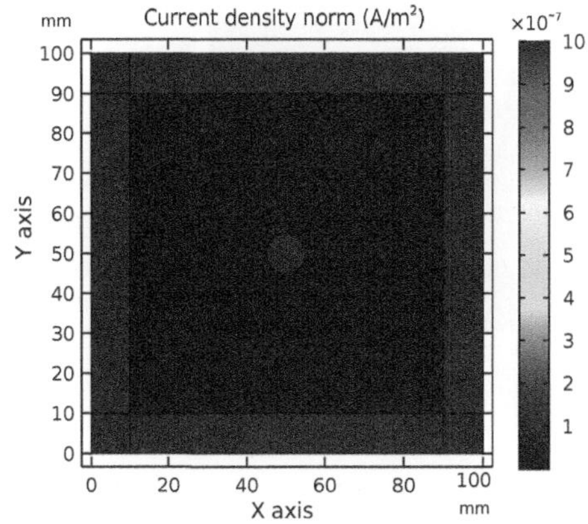

Fig. 4 Current density throughout RPC

4.2 Dark Current

The simulation of dark current generated upon the application of high voltage to the device has shown that the maximum current flows through the side and button spacers as the volume resistivity of the gas medium is high in comparison to that of the spacer material. The current density map inside the RPC of dimension 10 cm $\times$ 10 cm, has been depicted in Fig. 4. It can be found that the amount of dark current is governed by the resistivity of the spacer material as its value is higher in comparison to that of the electrodes [6]. It implies that the choice of spacer material is a crucial part of RPC construction.

4.3 Electric Field Configuration

It has been found from our study that instead of volume resistivity of either electrode or spacer material, the combination of these two parameters plays the governing role for the electric field. The electric field distribution inside the gas-gap with glass electrode and polycarbonate spacer is shown in Fig. 5(a) while that along X-direction (denoted by the red line in Fig. 5(b)) with different electrodes (bakelite and glass) and spacer (polycarbonate) combinations are shown in Fig. 5(c). The ratio of the volume resistivity of the spacer to the electrode should be around 10^2 or more to minimize the distortion in the electric field in the close vicinity of the side and button spacers. From the results, it can be inferred that the electrode resistivity ranging from 10^{10} to 10^{12} Ω cm, along with spacer resistivity 10^{14} to $10^{15}\Omega$ cm are the suitable choices of material for configuring an RPC with the given geometry.

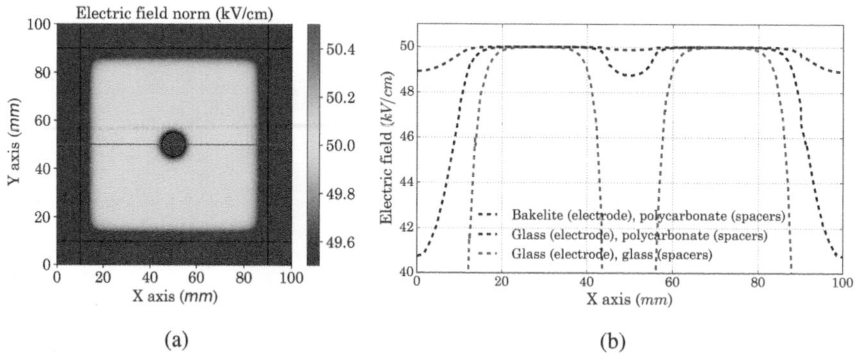

Fig. 5 (**a**) Electric field (kV/cm) map with glass (electrode) and polycarbonate (spacer), (**b**) Electric field value with different electrode and spacer configurations

5 Conclusions

The simulation of the voltage distribution, dark current and electric field configuration has provided us with a few criteria about the choice of materials in the construction of RPC. The non-uniform surface resistivity of the conductive coating between a range of 100 k$\Omega/\square$ to 1 M $\Omega/\square$ has been found to be suitable to maintain uniform distribution of high voltage on both the electrodes. It has been proved by the study that the total amount of dark current is governed by the resistivity of the spacer material. Higher volume resistivity of the spacer material to that of the electrode gives better electric field uniformity. In the future, we have plan to perform few experimental measurements to corroborate the simulation results.

Acknowledgements The authors, Subhendu Das and Jaydeep Datta, respectively acknowledge the support and cooperation of SINP, UGC, Govt. of India and INO Collaboration.

References

1. R. Cardarelli, V. Makeev, R. Santonico, NIMA **382**, 470 (1996)
2. B. Satyanarayana, Ph.D. thesis IIT Bombay (2009)
3. S. Ramo, Proc. IRE **27**, 584 (1939)
4. COMSOL Multiphysics Software. https://www.comsol.co.in/
5. S. Thoker, B. Satyanarayana, W. Bari, R. Shinde, G. Majumder, JINST **15**, T09010 (2020)
6. V. Ammosov, V. Korablev, V. Zaets, NIM **401**, 217 (1997)

Straw Tube Studies and Prototype Assembly for DUNE

Prachi Sharma, Bhumika Mehta, Riya Gaba, Vipin Bhatnagar, and Sushil S. Chauhan

1 Introduction

Deep Underground Neutrino Experiment (DUNE) [1], is an international particle physics experiment hosted by the Fermi National Accelerator Laboratory (Fermilab), U.S., that aims to answer some fundamental questions about our universe. The experiment comprises of three main components: (1) a highly intense neutrino source generated from a megawatt-class proton accelerator at Fermilab, (2) a Far Detector (FD) which is to be located about 1.5 km underground at the Sanford Underground Research (SURF) Facility in South Dakota, at a distance of 1300 km km from Fermilab, and, (3) a Near Detector (ND) that will be located just downstream of the neutrino source on the Fermilab site in Illinois.

The driving goals of the DUNE are to conduct a comprehensive program of neutrino oscillation measurements, search for proton decay in several decay modes, detect and measure the ν_e flux from a core-collapse supernova within our galaxy, if one happens during the lifetime of the experiment and other accelerator-based neutrino flavor transition measurements with sensitivity to BSM phenomena, measurements of neutrino oscillations using atmospheric neutrinos, searches for dark matter, and a rich program of neutrino interaction physics, including a wide range of measurements of neutrino cross sections and studies of nuclear effects.

The Long-Baseline Neutrino Facility (LBNF) beamline [2] will deliver the world's most intense neutrino beam to the near and far detectors in an on-axis configuration. The FD will be a modular LArTPC (Liquid Argon Time-Projection Chamber) with a total mass of 70 kt and a fiducial mass of roughly 40 kt. This detector will be able to uniquely reconstruct neutrino interactions with image-like precision and unprecedented resolution. The ND will include three primary detector components: a LArTPC, a high-pressure gaseous argon TPC (HPgTPC) surrounded by an

P. Sharma (✉) · B. Mehta · R. Gaba · V. Bhatnagar · S. S. Chauhan
Department of Physics, Panjab University, Chandigarh, India
e-mail: sharmaprachi55555@gmail.com

R. N. Patra (ed.), *Advanced Radiation Detector and Instrumentation in Nuclear and Particle Physics*, Springer Proceedings in Physics 282,
https://doi.org/10.1007/978-3-031-19268-5_9

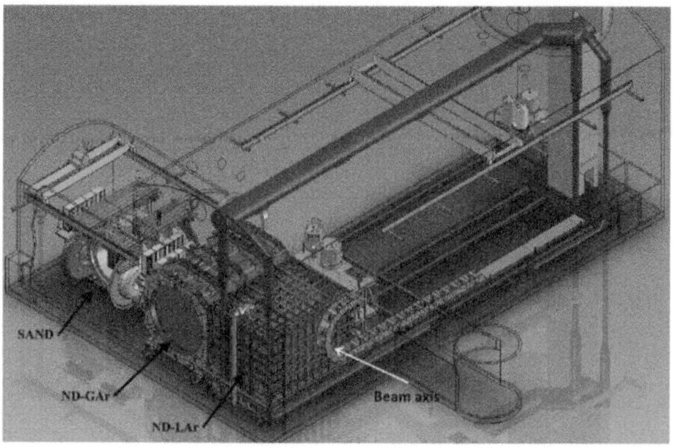

Fig. 1 Schematic of the DUNE ND hall shown with component detectors all in the on-axis configuration. The SAND detector is shown in position on the beam axis. Figure taken from [1]

electromagnetic calorimeter (ECAL) in a 0.5 T magnetic field, together called the multi-purpose detector (MPD) and an on-axis beam monitor called System for on-Axis Neutrino Detection (SAND). Figure 1 shows the DUNE ND Hall with component detectors.

SAND [3, 4] is a magnetized beam monitor that monitors the flux of neutrinos going to the FD from an on-axis position where it is much more sensitive to variations in the neutrino beam. SAND consists of an inner tracker surrounded by an ECAL inside a large solenoidal magnet. The inner tracker uses the Straw Tube as the tracking detector technology.

2 Straw Tube Tracker

Straw tubes are gaseous drift detectors that work in proportional operation mode. It is a cylindrical conducting tube filled with gas and a wire that is tensioned along the axis of the tube. The main purpose of the straw tube is to detect charged particles passing through it (mainly muons and pions). When a charged particle passes the tube, an electromagnetic interaction takes place between the charged particle and the atoms and molecules of the gas. Due to these Coulomb interactions, electron-ion pairs are created along the trajectory of the charged particle. The applied electric field between the wire (given positive voltage of few kV) and the tube results in the drifting of electrons and ions through the gas. The anode wire collects the electrons while the ions drift towards the straw tube wall (cathode). As the electric field is strong near the anode wire, the primary electrons formed drift to produce electron-ion pairs via the secondary ionization in the gas eventually forming an avalanche. This avalanche when reaches the anode wire is large enough to produce a measurable signal that can be recorded by the readout electronics. As the straw tubes work in the proportional region, the size of the signal is proportional to the deposited primary charge.

The tracking of the charged particles traversing the straw tube is done by the drift time measurement of the anode wire i.e. the arrival time of the signal defines the

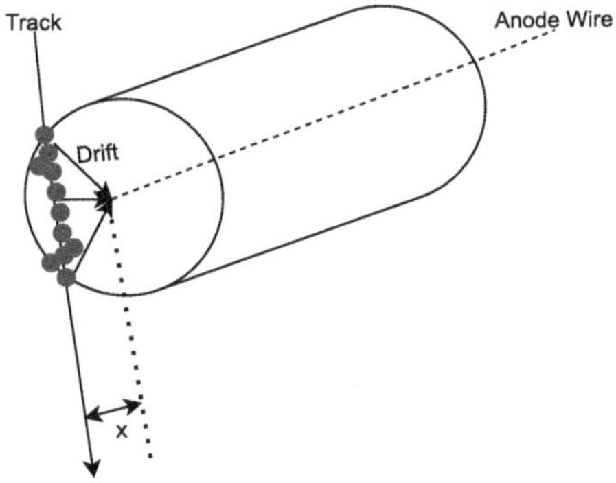

Fig. 2 Drawing of a straw tube for charged particle tracking. Distance of closest approach of charged particle track to the anode wire is denoted by x

drift radius, and the charge collected is proportional to the particle energy lost by ionization. The minimal distance between the wire and the trajectory of the charged particle is determined from the drift time of the ionized particles and this distance is the main information obtained using straw tube detectors for charged particles. Sketch of a straw tube is shown in Fig. 2.

3 Straw Tube Tracker for SAND

The Straw Tube Tracker (STT) is designed to be a fully tunable tracking detector providing control over the configuration, chemical composition and mass of the ν-targets, similar to electron scattering experiments [5].

The base tracking technology for the STT for SAND is provided by the low mass straws which have 5 mm diameter, 12 μm thick cathode walls and a 20 μm gold plated tungsten anode wire. The STT is to be operated with a gas mixture of Xe/CO_2 in the ratio of 70/30 and a pressure of 1.9 atm. The single hit space resolution for the straws is estimated to be <200 μm. As the targets are spread uniformly within tracker; the average density is low ($0.005 \leq \rho \leq 0.18$ g/cm^3) which helps to obtain the total detector length comparable to the radiation length, and thus an accurate measurement of the four-momenta of final state particles. Also, the excellent vertex, angular, momentum, and timing resolutions are the main factors to correctly associate neutrino interactions to each target material. Thus, SAND is a high resolution detector with momentum scale uncertainity < 2%. The thin replaceable targets account for 97% of the STT mass and the straws for the rest 3%.

4 STT *R & D* at Panjab University

Present Status
At Panjab University, we are operating a test ST Chamber that has been procured from JINR, Dubna, Russia. 50 straws have been purchased from Lamina Dielectrics Ltd., UK and a 20 μm gold plated tungsten wire was purchased from Luma Metall AB, Sweden. A leak test setup consisting of a gas sniffer, manometer, and pressure gauge has been established. A gas mixture Ar/CO_2 in the 80/20 volume ratio is available. An optical bench of dimensions $6' \times 4'$ has been made. A single channel preamplifier is also available which came with the test ST Chamber from JINR Dubna.

Work Done/In Progress: The assembly of a single straw tube has been done. A perspex gas chamber has been prepared for the assembled straw. Wire tension test has been done and the single straw tube performance study is ongoing (details in the next section). The proposed SAND STT will use 5 mm straws which will be procured from Lamina Dielectrics, UK.

1. Wire tension test and measurement: of the wire was measured with a Sonometer setup shown in Fig. 3. Resonant frequencies were observed for a particular mass and the masses were gradually increased. The properties of the wire used are shown in Table 1 and the results of the tension measurement of the wire with the sonometer setup is shown in Table 2.

Result: The 20 μm wire can hold a maximum tension of 60 g.

Fig. 3 Wire tension measurement setup at Panjab University

Table 1 Properties of the wire

Thickness	20 μm
Density	19.22 g/cc
Specific resistance	0.092 Ω-mm^2/m
Length (between 2 bridges)	75 cm

Table 2 Wire Tension Measurement with Sonometer

Mass (g)	Observed frequency (Hz)	Theoretical Values for Frequency (Hz)	Percentage Error (%) ((Theoretical Value-Observed Value)/(Theoretical Value))
20	124	120	3.33
30	145	147	1.3
40	162	169	4.14
50	188	189	0.5
60	201	204	1.4

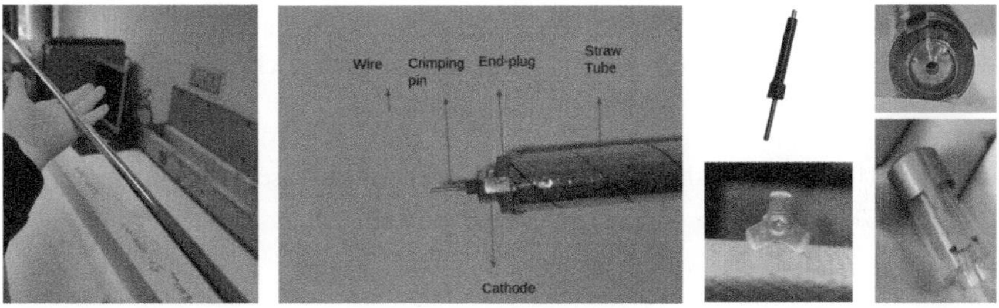

(a) (b) (c) (d)

Fig. 4 **a** Aluminum Coated Mylar Straw Tube (ST) **b** Assembled ST **c** Crimping pin (above) and Spacer (below) **d** Endplug

5 Assembly

The assembly of a single straw tube was done at Panjab University, Chandigarh. The Straw tube used is an aluminum coated Mylar tube. It is 181 cm long and has a diameter of 9.53 mm. A gold plated tungsten wire is used as anode and the cylindrical straw tube is the cathode. A spacer is placed in the middle of the straw tube and the wire is crimped with a crimping pin, after giving it the required tension. This setup is then glued to an endcap using Araldite glue leaving space for the passage of gas. Figure 4 shows the various components and the assembled ST.

The weight of the Straw Tube before assembly was 5.81 g and after the assembly was 9.76 g.

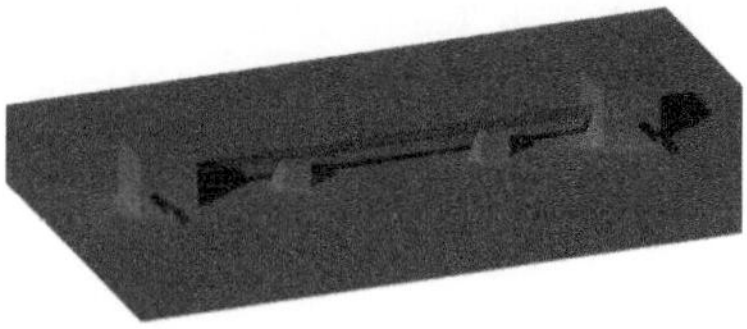

Fig. 5 (Right) Prototype designed for a single straw tube study

6 Future Plans

The group at Panjab University will work on the prototype shown in Fig. 5, which is being designed for the study of the assembled single straw tube. The gas chambers are to be fixed at the two ends of the straw tube and the straw tube holders (shown in yellow) under the ST. The inlet and the outlet of the gas in the perspex chambers are also shown. A PCB will be mounted at one side of the gas chamber. The gas to be passed will be a mixture of Ar/CO_2 in the ratio 80/20. The connection to the HV for the straw wire is shown in Fig. 6.

A prototype chamber shown in Fig. 7 has also been proposed with the dimensions of 1.8 m $\times$ 50 cm.

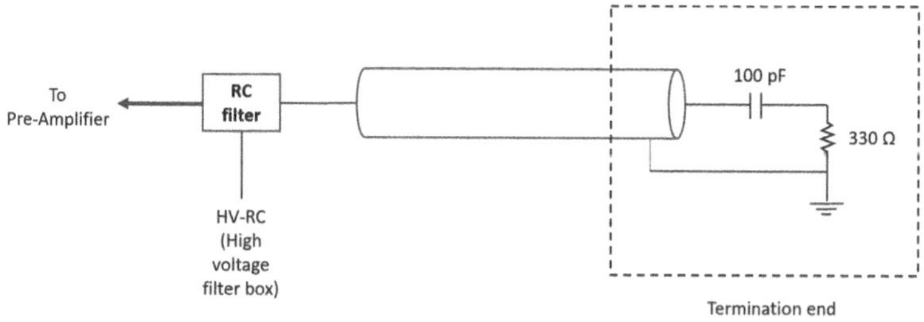

Fig. 6 (Right) A simplified schematic circuit of one detector channel for the Straw Tube

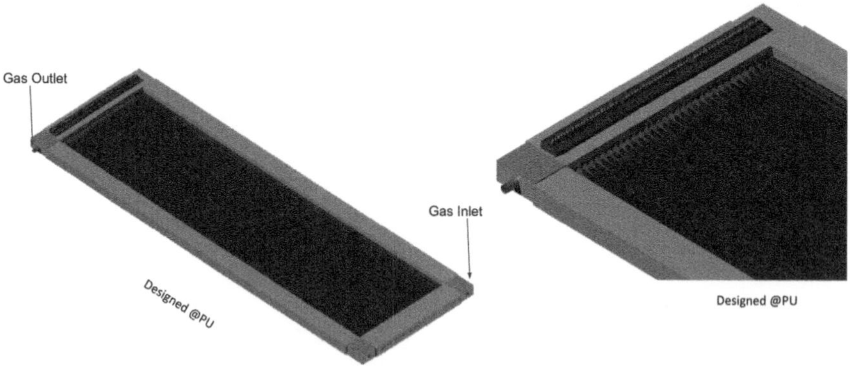

Fig. 7 3D CAD model of the proposed Prototype ST Chamber

References

1. B. Abi et al., Volume I. Introduction to DUNE. JINST **15**(08), T08008 (2020)
2. Acciarri, R. et al., Long-Baseline Neutrino Facility (LBNF) and Deep Underground Neutrino Experiment (DUNE) Conceptual Design Report Volume 1: The LBNF and DUNE Projects. https://arxiv.org/abs/1601.05471
3. R. Petti, Straw Tube Tracker for SAND: Design and Overview. https://indico.fnal.gov/event/43949/contributions/190339/attachments/130824/159732/DUNE-NDrev-09Jul20.pdf
4. R. Petti, Compact STT: Design and Performance (DUNE ND Workshop on Magnet Systems September 04, 2019). https://indico.fnal.gov/event/21272/contributions/61702/attachments/38592/46804/DUNEND-04Sep19.pdf
5. G.A. et al., A Proposal to Enhance the DUNE Near Detector Complex, DUNE doc 13262 (2019). https://docs.dunescience.org/cgi-bin/private/ShowDocument?docid=13262

Upgradation of CMS Detector at the LHC with GEM Detector

Harjot Kaur, Jyoti Babbar, Vipin Bhatnagar, Nitish Dhingra, Amandeep Kaur, Tanvi Sheokand, and J. B. Singh

1 Introduction

The Compact Muon Solenoid (CMS) [1] detector is a multipurpose detector installed at the Large Hadron Collider (LHC) [2] at CERN. CMS detector covers a broad-physics programme ranging from the precision measurement of the Standard Model (SM) processes to the search for the new physics phenomena. A powerful superconducting solenoid and an iron return yoke act as the basis and frame of the CMS detector. There are three sub-detectors which are located inside the solenoid: Tracker, Electromagnetic CALorimeter (ECAL) and the Hadron CALorimeter (HCAL). The muon system of the CMS is outside the superconducting magnet. Each sub-detector serves specific purposes, and provides an understanding of collision events. By the end of year 2023, the LHC is expected to reach the milestone of total integrated

H. Kaur (✉) · J. Babbar · V. Bhatnagar · A. Kaur · T. Sheokand · J. B. Singh
Department of Physics, Panjab University, Chandigarh 160014, India
e-mail: harjot.kaur@cern.ch

J. Babbar
e-mail: jyoti.babbar@cern.ch

V. Bhatnagar
e-mail: vipin.bhatnagar@cern.ch

A. Kaur
e-mail: a.kaur@cern.ch

T. Sheokand
e-mail: tanvi.sheokand@cern.ch

J. B. Singh
e-mail: jasbir.singh@cern.ch

N. Dhingra
Department of Soil Science, Punjab Agricultural University, Ludhiana 141004, India
e-mail: nitish.dhingra@cern.ch

R. N. Patra (ed.), *Advanced Radiation Detector and Instrumentation in Nuclear and Particle Physics*, Springer Proceedings in Physics 282,
https://doi.org/10.1007/978-3-031-19268-5_10

luminosity of 300 fb^{-1} including all the run periods. The high luminosity upgrade of the LHC is foreseen during a third long shutdown to further increase the instantaneous luminosity to 5×10^{-34} $cm^{-2}s^{-1}$. The high luminosity upgrade of LHC is aimed to exploit its full discovery potential in terms of the increased center-of-mass energy as well the instantaneous luminosity. This includes discovery of new massive particles and rare physics phenomena which otherwise are hard to access.

The muon system plays a crucial role in CMS as the production of new particles generally involves one or more muons. For example, one of the cleanest signatures for Higgs Boson search is the final state involving four muons from Higgs boson decay. The muon system of CMS detector consists of Drift Tubes (DT) in barrel region up to pseudo-rapidity $|\eta| < 1.2$, Cathode Strip Chambers (CSC) in the endcaps up to $0.9 < |\eta| < 2.4$ and Resistive Plate Chambers (RPC) provide redundant trigger and fine position measurement in both barrel and endcap regions. As the name signifies, detection of muons is one of the crucial tasks of CMS. The muon system aims to provide efficient and fast identification of muons in order to distinguish signal events from the background. However the forward region of the endcap is only instrumented with CSCs. The possible degradation of CSC performance due to the sustained operation in a high rate environment could drastically affect the entire muon system. However, major improvements are planned to cope up with the degradation of DTs, RPCs and CSCs by installing an additional set of muon detectors, called Gas Electron Multiplier (GEM) [3] in the first endcap muon station to improve and maintain the forward muon triggering and reconstruction in the region $1.6 < |\eta| < 2.2$ at high luminosity. The details on assembly and testing of these GEM detectors, as per the protocol set by CMS-GEM collaboration, are discussed in the subsequent sections of this paper.

2 GEM Detector

The GEM technology was introduced by Sauli [4] in the year 1997 to pre-amplify signals in Micro Strip Gas Counters (MSGCs). It consists of a thin layer of insulating polymer, usually a 50 μm thick polyimide, coated on both sides with 5 μm of copper and chemically perforated with a high density of microscopic holes. The typical diameter of the holes is 70 μm, with a pitch of 140 μm as shown in Fig. 1 (Left). The GEM detector works in a gas medium in which gas ionization takes place due to the acceleration of electrons under the influence of the electric field. These primary electrons acquire sufficient energy to produce an avalanche. The charge after the amplification can drift toward a readout board where it induces an electrical signal. Among noble gases, argon (Ar) has high specific ionization [6, 7] and is less expensive. However, it can cause avalanche creation beyond the Raether limit [4] which leads to sparks and cause permanent damage to the detector. Thus, carbon-dioxide (CO_2) is often used as quenching gas which can be used to reduce the spark probability and to stop unwanted avalanche due to the emission of UV photons. The gas mixtures of Ar and CO_2 are used in proportions of 70:30 for CMS GE1/1 detector.

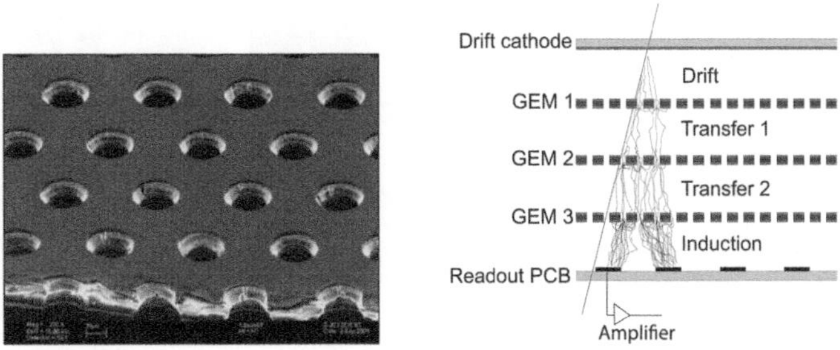

Fig. 1 Scanning Electron Microscope (SEM) picture of a GEM foil (Left) and Principle of operation of a triple GEM chamber (Right) [5]

The CMS triple GEM detector (GE1/1) is made by a drift cathode, three GEM foils, and the Printed Circuit Board (PCB) anode (or readout board). GE1/1 production chamber has a 3/1/2/1 mm (drift/ transfer1/ transfer2/ induction) gas gap configuration, as shown in Fig. 1 (Right), which is enclosed between two electrodes with an electric field applied between these electrodes. The primary electrons produce ionization in the gap between the cathode and the first GEM foil (drift gap) as shown in Fig. 1 (Right). A signal induced by a particle is not immediately re-absorbed by the GEM electrodes but is transferred through transfer region to another GEM foil for further amplification. The electron-ion pairs are attracted by electric field produced inside the GEM holes due to applied voltage across the two copper-clad surfaces of a foil. They acquire enough kinetic energy to produce secondary ionization in the gas and results in amplification of signal. After crossing the last GEM foil they drift through the induction gap to the anode where the signal is registered by the readout strips.

3 Assembly and Testing of GEM Detectors

3.1 Pre-assembly

A kit having different parts of the production chamber was made available by CERN. Prior to actual assembly of GEM detectors, some essential pre-assembly activities were carried out to make the raw material ready. As a first step, the screws and pullouts were cleaned using ultrasonic bath to remove the possible dust particles. In the next step, the drift board was prepared using the pullouts and by soldering the capacitors as well as resistors at their respective positions as shown in Fig. 2 (Left and Middle). Next step was to prepare the readout board by fixing the gas nozzles as shown in Fig. 2 (Right).

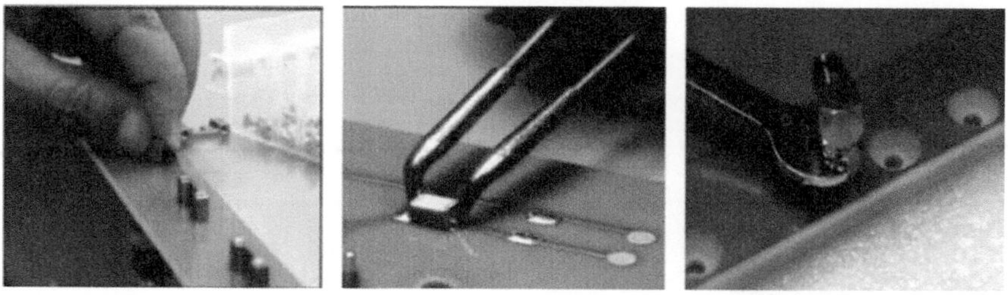

Fig. 2 Mounting of the pull-outs (Left), soldering of the SMD components (Middle) and fixing the gas nozzles (Right)

3.2 Assembly

The assembly of detector was done in a class 100 clean room built at Panjab University (PU) site. All foils were stretched in the frames for cleaning and testing. The foils were cleaned gently from both sides by using an anti-static sticky roller to remove the possible dust particles as shown in Fig. 3 (Left). Then foils were tested with the insulation meter by applying 550V voltage to foil for several minutes. As expected, the resistance of the foils was found to be 20 GΩ after few seconds with relative humidity lower than 40%.

The next step was the mounting of GEM stack. For this purpose, the plexiglass baseplate was placed on the assembly table and 3 mm clean spacers were inserted on the guiding pins. Afterwards, GEM1 foil (refer to Fig. 1 right) was placed having 1 mm spacers fixed over it as shown in Fig. 3 (Middle). Following the same approach, a stack was formed using GEM2 foil with 2 mm spacers and GEM3 foil with 1 mm spacers as shown in Fig. 3 (Right). The Plexiglass was again placed on the top of GEM foil stack using guiding pins and screws were fixed to protect the GEM stack. The excess of Kapton foils was removed using a sharp blade. The GEM stack was then dissociated from the plexiglass base plate and placed on the drift board in the area delimited by the pullouts as shown in Fig. 4 (Left). Then the chamber was moved to the assembly jig and drift board was clamped to the assembly table using aluminum bar. After placing the screws, the plexiglass and aluminium bars were removed. In the

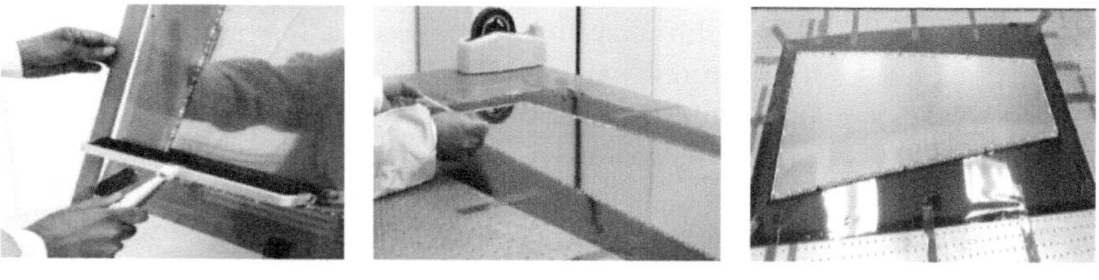

Fig. 3 Cleaning of the GEM Foils (Left), Placing the spacers (Middle) and Assembled GEM stack (Right)

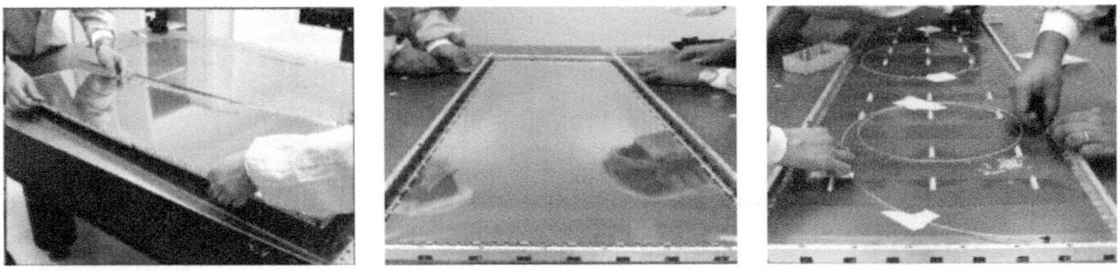

Fig. 4 Placing GEM stack on drift board (Left), Stretching the stack (Middle) and Assembled detector (Right)

next step, the foils were stretched to remove any wrinkles on the surface as shown in Fig. 4 (Middle). First the foils were stretched mechanically and later fine stretching was done using torque control tools through the pull-outs all along the foil edges, for a tension about 5 N/cm. Afterwards, the external frame was placed between the GEM stack and aluminum bars and readout board was put on the top of GEM stack as shown in Fig. 4 (Right). Finally, after tightening the screws at 1.2 Nm using the manual torque screw driver, the aluminum bars were removed to release the chamber.

3.3 Testing of GEM Detectors

After assembly, some quality control (QC) tests were performed [8]. QC2 test was performed before, during and after the assembly of the detector. After the assembly of GE1/1 detector, QC2 fast test was performed which aimed to determine the quality of a GEM foil by measuring the maximum leakage current in GEM foils and gas gaps. This test was carried out using Multi Mega-ohmmeter, also called Megger. The voltage of 550 V was applied on GEM top and bottom HV pads for several minutes, the impedance of the foils should reach 10 GΩ after few minutes. After the test was done, the GEM foils were discharged. Similarly, the impedance of the gaps was measured after applying voltage of 550 V on bottom HV pad of one GEM and top HV pad of other GEM foil. The detector was accepted if the impedance of all gaps was above 100 GΩ after a minutes. The next test in the series of checks was QC3 gas leak test which was meant to identify the gas leak rate of a detector by monitoring the drop of the internal over-pressure as a function of the time. All chambers were tested with an initial over-pressure of 25 mbar, and its internal pressure is monitored for 1 h. The maximum acceptable gas leak rate is about 7 mbar/hr and all the detectors assembled at PU site successfully passed the QC3 gas leak test because gas reduction rate was found to be less than 7 mbar/hr as shown in Fig. 5 (Left). To determine the V-I characteristics of a detector and identify possible malfunctions, defects in the HV circuit and spurious signals, QC4 test was performed. The monitored voltage (V_{mon}) as a function of the monitored current (I_{mon}) is shown in Fig. 5 (Right). The QC5 test is split into two steps: the measurement of the effective gain as a function of the

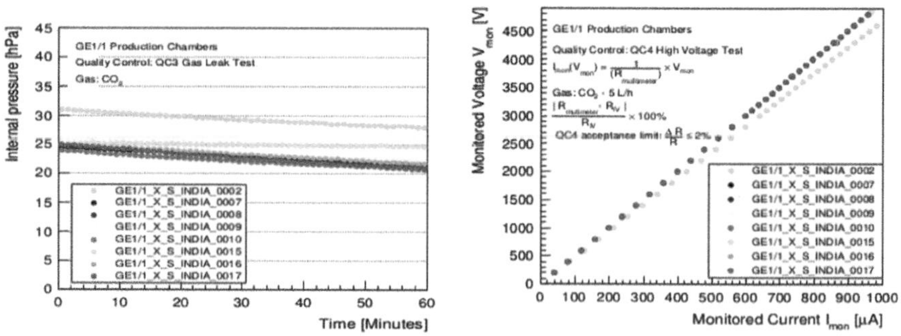

Fig. 5 Results of QC3 gas leak test (Left) and QC4 test (Right)

voltage applied on the divider; and the measurement of the response uniformity of the detector. This test was conducted at University of Delhi.

After performing the QC tests, the assembled chambers were shipped to CERN, Geneva. These chambers were equipped with the electronics such as VFATs, FEAST, optohybrid board, optical fibres, etc., [3]. A set of additional QC tests were performed for these chambers and the set of two chambers passing all the QC tests were coupled together to form one super chamber in order to obtain two detection planes. The super chambers alternate in the azimuthal direction between the long and the short versions to ensure the full coverage of the muon endcaps. These super chambers were later installed at CMS during year 2019–2020.

4 Phase-II Upgradation of CMS Detector

To extend the sensitivity for new physics searches, a major upgrade of the LHC known as High Luminosity LHC (HL-LHC) has been planned. With this the integrated luminosity will increase ten times with respect to the designed value. The center of mass energy of proton-proton collisions is expected to be raised from 13 TeV to 14 TeV. The high luminosity data taking period with the upgraded LHC, called Phase-II [9], is expected to end in 2038. To cope up with the increase in background rates and trigger requirements, two major upgrades known as GE2/1 and ME0 will take place in CMS muon station.

The GE2/1 upgrade, involves upgradation of second ring of GEM muon detectors in the endcap region next to ME2/1 chambers. The GE2/1 chambers will partially overlap with GE1/1 chambers and will cover the pseudo-rapidity range $1.62 < |\eta| < 2.43$. The main motivation behind the introduction of new ME0 detectors is to increase the geometrical acceptance for muons. The ME0 chambers will cover the pseudo-rapidity range $2.03 < |\eta| < 2.8$ using six layers of triple-GEM detectors.

The ME0 system will provide unique coverage in the range 2.4 < $|\eta|$ <2.8 and will strengthen the coverage provided by the CSCs, RPCs and GE2/1 in the range 2.03 < $|\eta|$ < 2.4.

5 Summary and Future Plans

The upgradation of CMS muon system is desired to improve and maintain the forward muon triggering as well as muon reconstruction at high luminosity. For this purpose, CMS detector is proposed to be equipped with an additional layer of GEM (GE1/1) detectors. Panjab University site was approved for the assembly and testing of these detectors by CMS-GEM group. The assembly and testing of 8 GE1/1 detectors was done and the experimental results were found to be consistent with the recommendations of CMS-GEM Collaboration. These detectors were eventually shipped to CERN, Geneva where they have been installed in the CMS detector successfully. For the phase-II upgrade of CMS detector, two sites from India will participate in the GE2/1 and ME0 production. The setup for QC5 test is now ready at Panjab University site, therefore, all QC tests upto QC5 for GE2/1 and ME0 chambers will be performed in near future.

Acknowledgements We are grateful to the Council of Scientific and Industrial Research (CSIR), Department of Science and Technology (DST), and University Grants Commission (UGC) for financial assistance. The CMS-GEM collaboration is also acknowledged for providing a kit having different components for GEM detector assembly and help during the assembly and QC testing.

References

1. CMS Collaboration, CMS Physics Technical Design Report, **5** (2006)
2. Evans et al., LHC machine. JINST **3**, S08001 (2008)
3. CMS Collaboration, CMS Technical Design Report for the Muon Endcap GEM Upgrade (2015) [CMS-TDR-013]
4. F. Sauli, The gas electron multiplier (GEM): operating principles and applications. Nucl. Instrum. Meth. A **805**, 2–24 (2016)
5. Gas Detectors Development Group. http://gdd.web.cern.ch/GDD
6. http://pdg.lbl.gov
7. http://cds.cern.ch/record/117989/files/CERN-77-09.pdf
8. J. Babbar et al., IOP Conf. Ser.: Mater. Sci. Eng. **1033**, 012055 (2021)
9. Mohit Gola and on behalf of CMS Muon Group, J. Phys.: Conf. Ser. **1498**, 012054 (2020)

FPGA Based High Speed DAQ Systems for HEP Experiments: Potential Challenges

Shuaib Ahmad Khan, Jubin Mitra, and Tapan K. Nayak

1 Introduction

Nuclear and particle physics experiments are performed to investigate fundamental interactions and the constituents of matter. In the framework of HEP experiment, the energies of the particle beam, constituted of either proton or heavy ions like lead ions, are of the order of hundreds of Giga-electron volts (GeV). Particle beams collide with each other to produce zillion of highly energetic particles. One of the best examples of such an experiment in the modern period is CERN's Large Hadron Collider (LHC).

In a heavily irradiated environment of a typical HEP experiment, various detectors are generally mounted. Depending on the physics rationale, each detector has a distinct form factor, geometry and granularity. Each high energy detector system has a significant number of electronics readout channels associated with it in order to retrieve information. With the growth in beam energies and luminosities with time, the rate and volume of data generated in a collider experiment has increased dramatically. Consequently, a large amount of raw data is generated at a rate of a few Tera Bytes per second (for example ∼4 Tera Bytes per second in ALICE experiment at CERN). An efficient DAQ and computing system is required to extract the data from the detectors, process the raw data in the front-end electronics (FEE), and transfer it reliably to the servers and the grid computing systems using the FPGA

S. Ahmad Khan (✉)
VECC, HBNI, Kolkata, India
e-mail: shuaib.ahmad.khan@cern.ch

J. Mitra
University of Oxford, Oxford, UK

T. K. Nayak
CERN, CH-1211 Geneva 23, Geneva, Switzerland
e-mail: tapan.nayak@cern.ch

R. N. Patra (ed.), *Advanced Radiation Detector and Instrumentation in Nuclear and Particle Physics*, Springer Proceedings in Physics 282,
https://doi.org/10.1007/978-3-031-19268-5_11

based processing units for the further offline computing. Traditional DAQ systems from the last century were framed around the features of specific experiments, with custom designed readouts having only hundred of readout channels and non-standard interconnects acknowledging DAQ concerns. Also these are prone to multi-bit upset in radiation zone.

In the recent times, the availability of cutting-edge technology such as densely packed FPGAs and graphics processing units (GPU) as commercial-off-the-shelf (COTS) components have pushed the data handling technologies and topologies to their limits. With the advent of point-to-point high-speed networks such as Ethernet and the Peripheral Component Interconnect Express (PCIe) protocol; data of the order of Terabytes/sec, high channel counts, and on-board and local processing are achieved.

We have presented an overview of the new FPGA based high speed DAQ readout scheme and its advantages over the conventional approach in Sect. 2. Potential technical challenges in the development of such a system, points of uncertainties and their probable solutions are summarized in Sect. 3. The results with a deep analysis for the proper selection of FPGAs are discussed in Sect. 4. The paper is concluded with a summary in Sect. 5.

2 Data Acquisition for HEP Experiments and the Role of FPGAs

The design and development of acquisition systems for HEP experiments is a complex process. The detectors are positioned in the radiation hard zone with FEE mounted over detectors. To transport the signal across greater distances, FEE turns the raw detector charge to an electrical signal. The requirement for radiation resistance for the cavern's installations makes DAQ's design a major consideration. To get around this, a design topology is generally adopted in which application specific integrated circuits (ASICs) are used in the stringent radiation zone and FPGA-based circuits are used for the rest of the DAQ chain development. FEE are made using custom designed ASICs. Data from the detector backplane is sent to the FEE. The next stage is to deploy dedicated FPGA-based data concentrator units (CU) that are placed in non-radiation hard counting rooms as indicated in Fig. 1. CUs combine data from numerous channels and multiplex it onto a smaller number of high-bandwidth channels. The data is processed by CU before being transferred to the servers via high-speed links. FPGA based CUs are also used for data processing in real time. Processing in real time aids the rejection of unnecessary data prior to its storage. The use of CU for online processing necessitates the careful selection of an on-board FPGA with the necessary specifications.

The majority of LHC experiments are presently operated in this intelligent manner. This topology reduces the time and effort necessary for development while increasing performance. Many essential design aspects are taken care of with the usage of

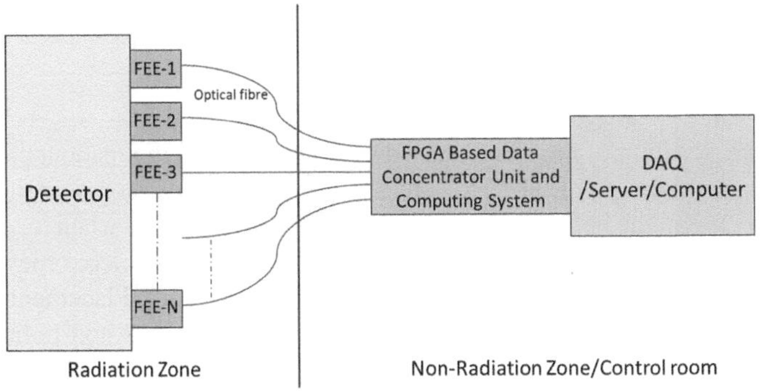

Fig. 1 Block diagram: Scheme of DAQ for HEP experiments

FPGAs. Space, cooling, magnetic fields, ionising radiations, power limits, high data rate handling in FPGA, low and fixed latency transmission, and data loss recovery in radiation situations against multibit upsets[1] are only a few examples. Because COTS components are used, the DAQ has a high availability and is easy to maintain during operation. This is an important metric to consider when installing DAQ systems. Some detectors provide data at a faster rate and in greater volume than others, necessitating the use of a reconfigurable DAQ system to balance the workload distribution. Online data reduction also requires on-the-fly data processing, with only the filtered data being saved for fine analysis. The use of FPGAs aids in meeting these important requirements. An insightful design for DAQ necessitates a well-researched and explored approach to selecting the FPGA, which will be covered in the following sections.

3 Potential Challenges

The FPGA based CUs are at the core of the DAQ, passing the trigger, timing, and data. Consequently, these boards must be extremely reliable. Failure of CUs during the run time of an experiment will result in the loss of beam data before being replaced, which is a highly undesirable situation given that the LHC beam is the result of several intensive efforts at the machine level. Technical challenges in the design and development of CUs are summarised as follows:

[1] Multibit upset: if two or more error bits occur in the same data word.

3.1 Location of the FPGA Based CUs and Selection of FPGAs

The physical location of the FPGA based readout boards in the readout chain affect the selection and design of the hardware. In contrast to its location near the detector systems in cavern, the boards will be accessible during operation when located in the ground level counting room. Also there will be no additional electronics, cooling, cabling, to be installed and maintained along with the detector. Placement of FPGA boards in the non-radiation hard zone allows the use of various non-radiation hard FPGAs available commercially. A detailed analysis focussing on the chief performance metrics of FPGAs is performed on the available devices from the popular vendors of equal repute like Intel Inc. and Xilinx Inc. and the results are shown in Sect. 4.

3.2 Signal Integrity Issues

In a widely distributed readout scheme of HEP experiments, there are multiple points of uncertainities which leads to the loss of synchronization. Maintaining the timing relationships and synchronization among payloads is an essential pre-requisite for the integrity of the acquired data. When signals are exchanged across asynchronous clock domains, metastability may be introduced [1]. The register setup and hold time should be kept constant to ensure data read and write reliability. Furthermore the temperature variations could onset metastable behaviour that accounts for the phase drifts of the clocks and push the system in an unstable zone.

Eye diagram analysis and jitter measurements are used as metrics to indicate the signal integrity of the system [3]. It acts as a measure for the performance of the link as shown in results.

3.3 Hardware Complexities

The custom designed FPGA boards for the high speed data transmission requirements of the HEP applications are highly complex with multiple optical transceiver links [2]. Transmission loss in the data is seen if not properly designed at hardware level. The boards are fabricated on the multilayered PCBs with requirement of laser drilling to match the via-size requirements and specialized soldering techniques are also needed. Material requirements for the PCBs are also strict as the boards are exposed to high thermal loads. Hence PCB material should withstand high Glass Transition

Temperature (Tg)[2] and high Decomposition temperature (Td),[3] as a measure for the degradation of the material. Td is the point as which reliability is compromised and delamination may occur. Data will be aggregated to even higher speed PCIe Gen3 $\times$ 16 links hence low dielectric constant material is required to reduce the high frequency losses at high data rates.

4 Results and Discussion

Various considerations constrain the FPGA variant for DAQ hardware development and firmware design in the topology when FPGAs are kept outside the radiation zone. State-of-the-art non-radiation hard FPGA families from Xilinx and Intel inc. are compared against numerous essential characteristics as mentioned in the Table 1. It provides an in-depth analysis of the metrics [3] like process technology which gives an idea of the length of a transistor gate: the smaller the gate size the more processing capability that can be packed into a given space channel, availability of High Speed Serial Interface Serializer-Deserializer (HSSI-SerDes) like 10 Gbps transceivers on FPGA, amount of logic resources, previous proven use, market research, Phase lock loops (PLL), and their availability on the market. These factors influence the FPGA chip that is used on the CU and the computing units. The survey is quantitative and derived from the device configuration and the manufacturer specification. The critical requisites of Arria-10, Agilex-F, Stratix-V GX, Stratix-10 FPGAs of Intel inc. with Virtex-7, Virtex-6, Virtex Ultrascale FPGAs of Xilinx inc. are compared. According to the results shown in Table 1, the latest 10-nm Intel's Agilex-F FPGA and 20-nm Xilinx's Virtex Ultrascale FPGA have 2.69 million Logic Elements/cells and 1.9 million Logic Elements/cells respectively, and are most powerful in terms of processing and logic resources. Both devices also support the latest generations of PCIe and 10Gbps transceivers. The designer's optimal device selection, on the other hand, is influenced by the experiment's processing requirements. For design engineers, the results presented are an in-depth but not exhaustive survey in tabular form and serve as a valuable reference. The statistics of the eye diagram immediately reveal the signal to noise ratio of the high-speed data transmission. The eye width is determined by the jitter in the transmission. As an example; the signal quality of the 4.8 Gbps rad-hard GBT protocol [3] on Stratix-V is measured using the eye diagram, on a high bandwidth data analyser oscilloscope as shown in Table 2. Jitter is of the order of pico-second only.

[2] Tg: temperature range in which a PCB substrate transitions from a rigid state to a deformable state; however reversible.

[3] Td: the temperature at which a PCB material chemically decomposes and not reversible.

Table 1 FPGA selection parameters

FPGA Family Name	Intel Agilex-F	Intel Stratix-V GX	Intel Stratix-10	Intel Arria-10 GX	Xilinx Virtex-6	Xilinx Virtex-7	Xilinx Virtex Ultrascale
Status	Available	Available	Available	Available	Available	Available	Available
FPGA part number		5SGXEA7	10SG28	10AX115	XC6VLX240T	XC7VX690T	XCVUI90
Process technology	10 nm	28 nm	14 nm	20 nm	40 nm	28 nm	20 nm
PLLs		28	48	32	12	20	60
>=**10Gb/s Transceivers**	64xNRZ	48	144	96	24	80	60
	48xPAM-4						
Logic Elements/cells[M]	2.69	0.622	2.8	1.15	0.241	0.693	1.9
Look up Table (LUTs)[M]	0.912	0.235	1.8	0.425	0.15	0.433	1.07
Flip Flops (FFs)[M]	11.8	0.939	7.4	1.7	0.3	0.866	2.14
18/20Kb RAM Blocks	13272	2560	11721	2713	832	2940	7560
PCIe x8,Gen3	upto Gen5	4	6	4	2(Gen2)	3	6
Total Block RAM(Mb)	259	50	229	53	15	53	133
Used to develop (Name of card)		AMC40 card		PCIe40 card	C-RORC board	MP7 card	CRI board

Table 2 Eye Diagram for data of GBT transmission

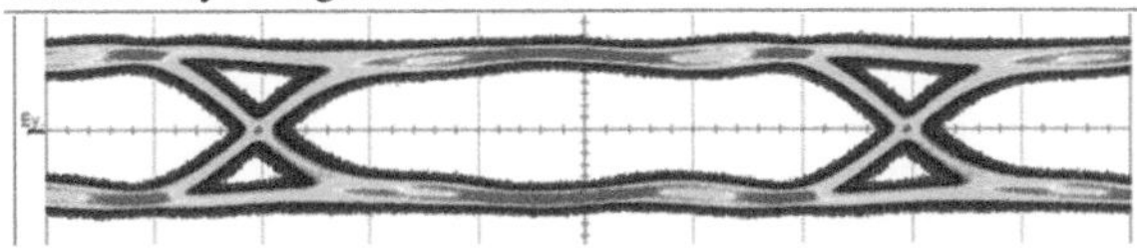

Parameter	Value
Eye width	176.8 ps
Eye height	373 mV

5 Summary

An overview of the new FPGA based high speed DAQ system which is capable of high data rate communication is presented. Technical challenges in its development are also highlighted which includes the placement of FPGA boards in the readout chain, signal integrity issues and hardware complexities in the fabrication. A thorough examination of FPGA parameters for optimal device selection is presented.

References

1. S.A. Khan et al., Development of FPGA based phase alignment logic for the high speed protocol in HEP Experiments. J. Comput. Phys. Commun. **259** (2021)
2. J. Mitra et al., GBT link testing and performance measurement on PCIe40 and AMC40 custom design FPGA boards. J. Instrum. **11**, C03021 (2016)
3. J. Mitra et al., Common Readout Unit (CRU)-A new readout architecture for the ALICE experiment. J. Instrum. **11**, C03039 (2016)
4. Andina et al., *FPGAs: Fundamentals, Advanced Features, and Applications in Industrial Electronics* (CRC Press, 2017)

Gain Uniformity of a Quad-GEM Detector at Different Gas Flowrates

Rupamoy Bhattacharyya, Rama Prasad Adak, Pradip Kumar Sahu, and Sanjib Kumar Sahu

1 Introduction

The Gas Electron Multiplier (GEM) detector [1] is widely used in many high energy physics experiments [2–4]. A 10 cm × 10 cm prototype quad-GEM detector, fabricated at the Institute of Physics, Bhubaneswar [5, 6] is utilized for this purpose. The drift, transfer and induction gap of the quad-GEM detector are 3, 2 and 2 mm respectively. Detector gain and gain uniformity across the detector are the crucial parameters for detector characterisation. Both of them are systematically studied at different gas flowrates. The gain is calculated from the measured anode current of the detector using a radioactive Fe^{55} X-ray source. the absolute gain (G) of the quad-GEM detector is calculated using the following relation [7]

$$G = \frac{I}{Rne}. \quad (1)$$

Here I is the anode current, R is the count rate of the radioactive source, n is the number of primary electrons generated inside the drift region for each incoming ionizing particle and e is the electronic charge.

R. Bhattacharyya (✉) · P. K. Sahu · S. K. Sahu
Institute of Physics, Bhubaneswar, Odisha, India
e-mail: rupamoy@gmail.com

R. P. Adak
Department of Physics, Taki Government College, Taki, West Bengal, India

P. K. Sahu · S. K. Sahu
Homi Bhabha National Institute, Mumbai, India

R. N. Patra (ed.), *Advanced Radiation Detector and Instrumentation in Nuclear and Particle Physics*, Springer Proceedings in Physics 282,
https://doi.org/10.1007/978-3-031-19268-5_12

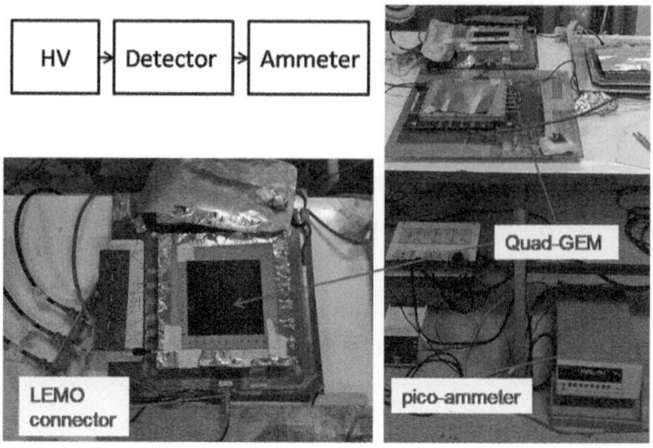

Fig. 1 Left panel: Block diagram of the experimental setup (top), top view of quad-GEM detector (bottom); Right panel: Experimental setup for measuring detector gain

2 Experimental Setup

The quad-GEM detector is biased using a four-channel programmable high voltage power supply (CAEN N1470). A pre-mixed gas mixture of Ar:CO_2 in the ratio of 80:20 is used during the experiment and the gas flowrate is monitored continuously using a flow-sensor built in-house [8]. A constant $\Delta V_{GEM} = 300$ V is maintained across each GEM foil. The corresponding electric field in the drift, transfer and induction gaps are 0.4 kV/cm, 3.6 kV/cm, and 4 kV/cm respectively. The surface area of the detector ($10 \times 10\,cm^2$) is divided into 16 equal zones of area $2.5 \times 2.5\,cm^2$ and each zone is irradiated using a collimated Fe^{55} X-ray source [9]. The corresponding anode current is measured using a pico-ammeter (Keithley 6485). The block diagram and experimental setup are shown in Fig. 1.

3 Results and Discussions

2D distribution of anode current and absolute gain across the 16 zones of the detector for flowrates 5.3, 7.1, 10.2, 15.1, 21.5 and 26.9 SCCM are shown in Figs. 2, 3, 4, 5, 6 and 7 left and right panel, respectively. The background current is limited to 0.3 nA in each of the cases. The present count rate R is measured using a pre-amplifier (Ortec 142AH), spectroscopic amplifier (CAEN N968) followed by a discriminator and scaler module (built in-house) [7] and it is found to be 9 kHz. Results show that the gain is relatively less along the sides (especially at the corners). A part of the avalanche of electrons, might not be collected by the anode planes in these regions. Variation of average absolute gain vs. flowrate is shown in Fig. 8. In each of the cases, the gas flow direction inside the chamber is from bottom-left to top-right. Next, for a specific flowrate (21.5 SCCM) the gas flow direction inside the gas chamber is reversed (top-right to bottom left) and the corresponding anode current and gain distribution is shown in Fig. 9. After calculating the average gain for each zone from

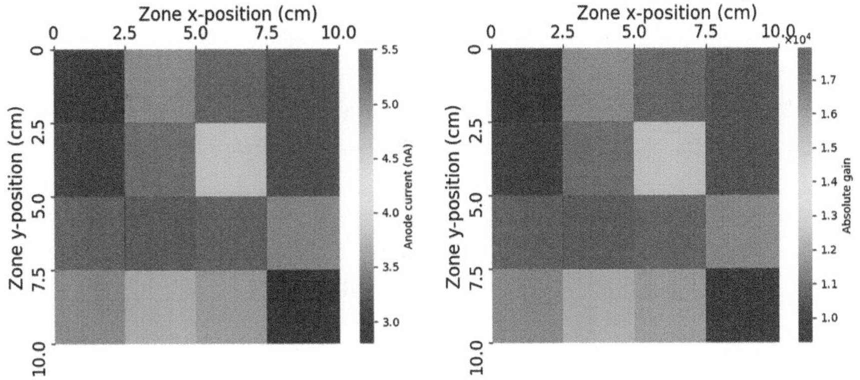

Fig. 2 2D distribution of anode current (left panel) and 2D distribution of gain (right panel) at a flowrate of 5.3 SCCM

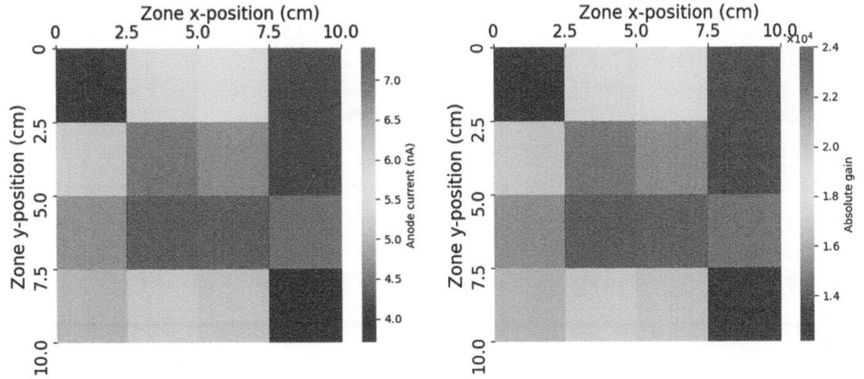

Fig. 3 2D distribution of anode current (left panel) and 2D distribution of gain (right panel) at a flowrate of 7.1 SCCM

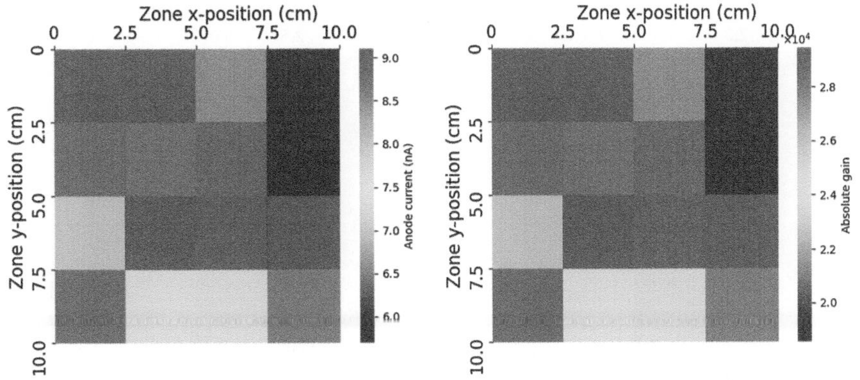

Fig. 4 2D distribution of anode current (left panel) and 2D distribution of gain (right panel) at a flowrate of 10.2 SCCM

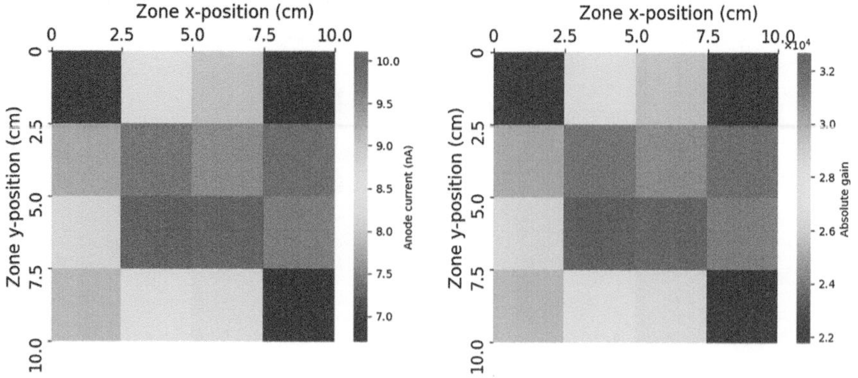

Fig. 5 2D distribution of anode current (left panel) and 2D distribution of gain (right panel) at a flowrate of 15.1 SCCM

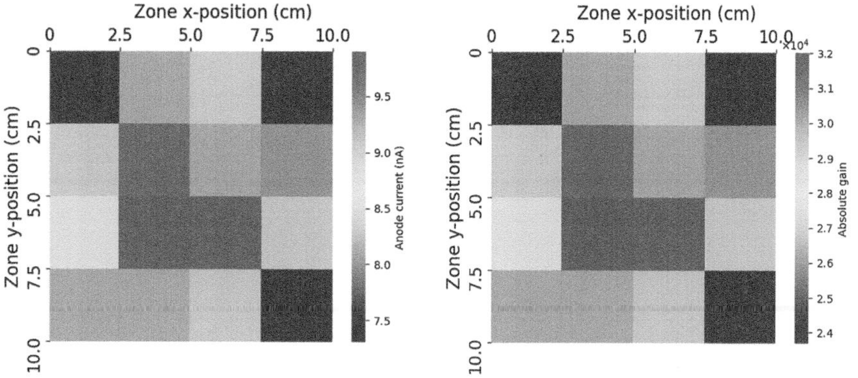

Fig. 6 2D distribution of anode current (left panel) and 2D distribution of gain (right panel) at a flowrate of 21.5 SCCM

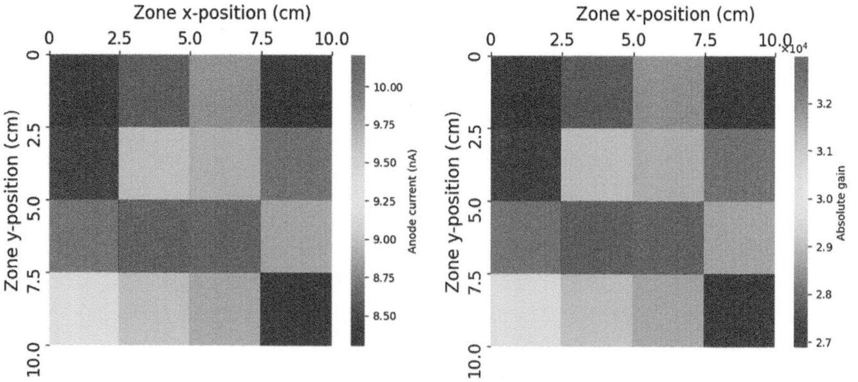

Fig. 7 2D distribution of anode current (left panel) and 2D distribution of gain (right panel) at a flowrate of 26.9 SCCM after reversing the flow direction

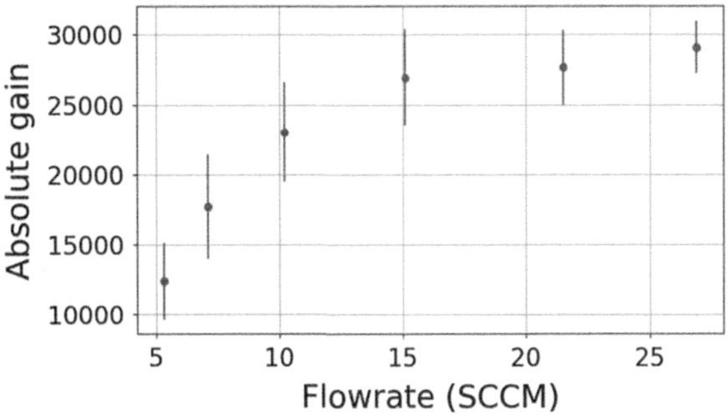

Fig. 8 Variation of absolute gain verses flowrate at a constant $\Delta V_{GEM} = 300$ V

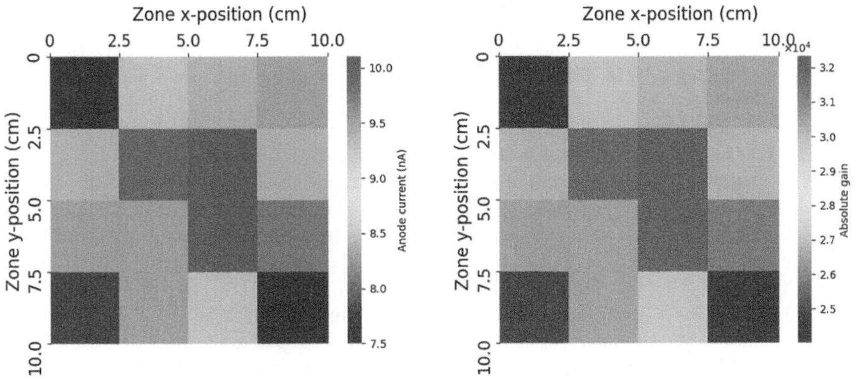

Fig. 9 2D distribution of anode current (left panel) and 2D distribution of gain (right panel) at a flowrate of 21.5 SCCM after reversing the flow direction

Figs. 6 and 9 (right panel), it is found that there is a 9.7% variation of gain across the detector for a fixed flowrate of 21.5 SCCM at a constant $\Delta V_{GEM} = 300$ V.

4 Conclusion

The variation of absolute gain with gas flowrate is studied for a prototype quad-GEM detector fabricated at the Institute of Physics, Bhubaneswar, for a fixed $\Delta V_{GEM} =$ 300 V. The absolute gain of the detector initially increases with the gas flowrate (as the number density of the ionizing gas increases with flowrate) and tends to saturate from 15 SCCM (which sets the operating flowrate for this detector). So, flowrate of the gas plays a crucial role in determining the gain of the quad-GEM detector, especially for low ($<15 SCCM$) flowrates. The gain uniformity is studied for a specific flowrate (by reversing the flow direction as well) and it is found that there is a spread of 9.7% over the 16 zones of the active area of the detector in the experimental setup. It may be mentioned here that the presence of pressure gradient near the gas inlet and outlet regions or/and the in-homogeneity of the gap between the gem foils (if any) could result in such variation of gain across the detector.

References

1. F. Sauli, GEM: a new concept for electron amplification in gas detectors. Nucl. Instrum. Methods A **386**, 531–534 (1997). https://doi.org/10.1016/S0168-9002(96)01172-2
2. ALICE TPC collaboration, The upgrade of the ALICE TPC with GEMs and continuous readout. J. Instrum. **16**, P03022. https://iopscience.iop.org/article/10.1088/1748-0221/16/03/P03022
3. A. Colaleo, et al., CMS technical design report for the muon endcap GEM upgrade (2015). https://cds.cern.ch/record/2021453, CERN-LHCC-2015-012, CMS-TDR-013
4. A.K. Dubey et al., GEM detector development for CBM experiment at FAIR. Nucl. Instrum. Methods A **718**, 418–420 (2013). https://www.sciencedirect.com/science/article/abs/pii/S0168900212011783
5. S. Swain et al., A quad-GEM detector prototype operated at very low gas gain. J. Instrum. **12**, T07002 (2017). https://doi.org/10.1088/1748-0221/12/07/T07002
6. A. Tripathy et al., Measurement of ion backflow fraction in GEM detectors. Nucl. Instrum. Methods A **1013**, 165596 (2021). https://doi.org/10.1016/j.nima.2021.165596
7. R. Bhattacharyya et al., Remote data acquisition with wireless communication for a quad-GEM detector. J. Instrum. **17**, T02001 (2022). https://doi.org/10.1088/1748-0221/17/02/T02001
8. S. Sahu et al., Building of gas flow monitor for GEM detector. DAE Symp. Nucl. Phys. **61**, 1002–1003 (2016). http://www.sympnp.org/proceedings/
9. R.N. Patra et al., Measurement of basic characteristics and gain uniformity of a triple GEM detector. Nucl. Instrum. Methods A **862**, 25 (2017). https://doi.org/10.1016/j.nima.2017.05.011

Characterization of Sapphire Detector for CEνNS Search at MINER

Mouli Chaudhuri

1 Motivation

Coherent neutrino-nucleus scattering (CEνNS) is a Standard model process where a neutrino interacts with a nucleus as a whole by exchanging a Z boson and scatters off the nucleus creating a nuclear recoil [1]. CEνNS requires incident neutrino energies of the order of a few MeV which results in a nuclear recoil energy of 1 eV to few keV depending on the mass of the target nucleus [2]. MINER experiment at the Nuclear Science Center (NSC) at Texas A&M University, USA aims to measure precisely the CEνNS cross-section utilizing the large neutrino flux from the 1 MW reactor with ^{235}U core. In Table 1 some details of the MINER reactor are listed. Figure 1a shows the neutrino spectrum from a reactor [3] where the blue dotted vertical line in the figure represents energy threshold for Inverse Beta Decay (IBD) process. As most of the neutrino flux is below this threshold, they could be detected through the CEνNS process. Detection of low energy neutrino demands low recoil energy threshold detectors ($\sim$100 eV). Figure 1b shows the maximum recoil energy as a function of neutrino energy for various target masses (solid lines). It is seen that reactor neutrinos impart higher recoil energies to low mass nuclei like Al (red solid line) and O (black solid line). Standard model predicts that in coherent scattering process the interaction cross-section is enhanced by a factor of N^2, where N is the number of neutrons in the target nucleus as opposed to IBD [2]. The interaction cross-section as a function of neutrino energy (dash-dotted lines) is shown in the

on behalf of MINER collaboration

M. Chaudhuri (✉)
School of Physical Sciences, National Institute of Science Education and Research, Jatni 752050, India

Homi Bhabha National Institute, Training School Complex, Anushaktinagar, Mumbai 400094, India
e-mail: mouli.chaudhuri@niser.ac.in

R. N. Patra (ed.), *Advanced Radiation Detector and Instrumentation in Nuclear and Particle Physics*, Springer Proceedings in Physics 282,
https://doi.org/10.1007/978-3-031-19268-5_13

Table 1 MINER reactor details

Reactor fuel	^{235}U (20%, LEU)
Reactor power	1 MW
Energy per ^{235}U fission	200 MeV
Neutrino yield per fission	∼6
Neutrino energy per fission	1.5 MeV
Neutrino flux at 1 m from core	∼10 cm^{-2}s^{-1}

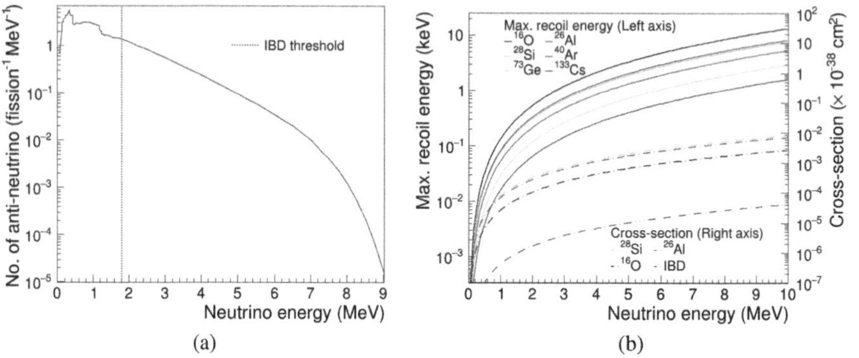

Fig. 1 **a** Reactor neutrino spectrum where X axis is the neutrino energy in MeV and Y axis is the anti-neutrino flux per fission per MeV. The blue dotted line shows the threshold for Inverse Beta Decay (IBD), **b** Maximum recoil energy in keV versus neutrino energy in MeV (left axis) where Al and O show higher recoil energy and cross-section versus neutrino energy plot (right axis) where the blue dash-dotted line shows the cross-section for IBD and upper curves show for the Al, O and Si through CEνNS which is two order magnitude higher than IBD

same Fig. 1b plotted in right Y axis. Hence, a detector madeup of Al and O would be very good for CEνNS search using reactor neutrinos. A sapphire scintillation detector of 100 g is fabricated and characterized at Texas to measure CEνNS. Here we report the detector performance at NSC with reactor off condition.

2 Detectors and Its Detection Principle

Sapphire is a scintillating crystal. Phonons and light/photons are generated simultaneously from an particle interaction. The detector has a diameter of 76 mm, thickness of 4 mm and has a mass of 100 g. The phonons are collected by the Transition-Edge-Sensors (TES) photo-lithographically placed on the surface of the detector. Figure 2 shows a picture of the sapphire detector where the phonon sensors consists of ∼1000 TES, divided into 4 groups of ∼250 sensors forming independent readout channels, A, B, C and D. This type of configuration helps to reconstruct the interaction posi-

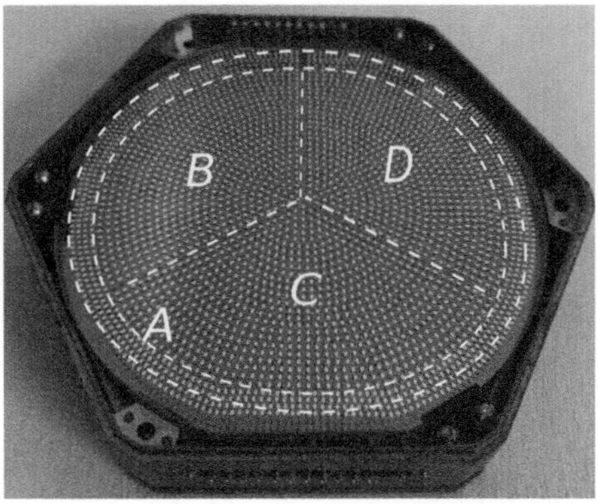

Fig. 2 Sapphire scintillation detector in Cu casing showing 4 independent readout channels, A, B, C and D where B, C, D are inner channels and A is outer

tion based on the relative amplitudes between the channels. The TES consists of two superconducting materials Aluminium (Al) and Tungsten (W). The W layer is on the top of the Al layer. The detector is operated at a cryogenic temperature of ~50 mK to keep the Al and W at their superconducting phase.

When the phonons reach the surface of the sapphire crystal Al fins absorb the phonons. These phonons will break the cooper pairs within the Al and create quasi particles and diffuse into W layer which has a lower superconducting bandgap energy compared to Al. Thus, phonons get trapped in the W layer that heat up the W from superconducting to normal state. This will create a sharp change in the resistance which can be detected using SQUIDs (superconducting quantum interference device). Sapphire detector has high phonon collection efficiency due to better matching of Al phonon collector fins to Al_2O_3 substrate.

3 Experimental Setup

The top-down view of the 1 MW pool type reactor at NSC has been shown in Fig. 3a. The bright light seen in the figure is the reactor core and the white outline represents the proposed experimental setup with shielding. The reactor is surrounded by a high density concrete wall. In the recent MINER engineering run, the sapphire detector together with two cryogenic germanium detectors was placed at ~4.5 m from the reactor core inside a dilution fridge to operate at cryogenic temperature (~50 mK). The germanium detectors have the similar configuration like the sapphire detector, 76 mm in diameter and 4 mm in thickness with 4 readout channels (outer A and B, C, D inner channels). The fridge was surrounded by lead and water bricks as shown in Fig. 3b to shield high energy gammas and fast neutrons. The schematic of the detector tower inside the dilution fridge in MINER site is shown in Fig. 3c. The sapphire detector was sandwiched between two germanium detectors of same size

(a) (b) (c)

Fig. 3 **a** The top-down view of the reactor at NSC where the proposed shielding for the experiment is shown by the white outlines. **b** shows the experimental setup during recent engineering run at MINER with full Lead and water bricks shielding. **c** Schematic of the detector tower at MINER

for single scatter background rate calculations as most of the rare events are single scatter events. They rarely interact with the detector material whereas backgrounds are multiple scatter events (Explained in detail in Sect. 4.1).

4 Data Analysis and Results

The data is taken with a 14 channel CAEN digitizer. The readout channels for sapphire detector are A, B, C and D whereas the readout channels for both the germanium detectors are A and C. The raw data is stored as pulses in the form of ADC units as a function of time. Noise data is taken separately with a random trigger. The pulse shape can be characterized by its pre and post pulse baselines, peak, risetime and falltime parameters. By putting appropriate cuts on these parameter values, we select good pulses and by averaging over all these good pulses a pulse template is made. From the noise data, we obtain the Power Spectral Density (PSD). The noise PSD, the pulse template and the raw data, all act as input to a method called as 'Optimal Filter' (OF) [4]. In the optimal filter method the pulse template is fit to each raw pulse in the data. The fitting is done in the frequency domain by taking a Fourier transform of the pulse and the template. The amplitude of the pulse is determined from the best fit result. The OF amplitude distribution of the entire data set can then be calibrated to a source with known energy.

4.1 Results from Experimental Site at MINER

The detector is made of Al_2O_3. So, Al and O could be in-situ backgrounds for this detector as they show fluorescence. Also Cu could be another background candidate

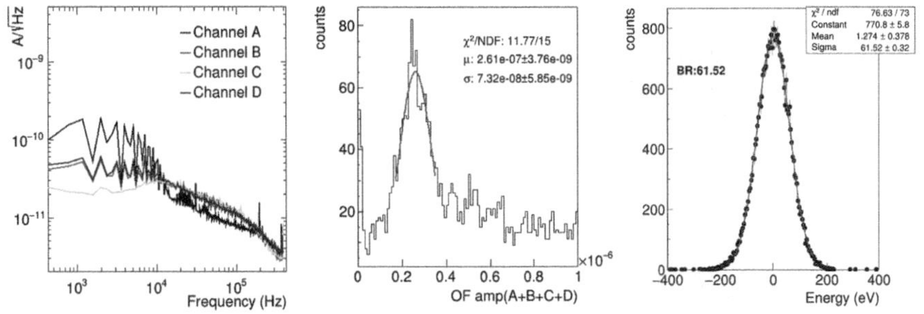

(a) (b) (c)

Fig. 4 **a** Noise PSD of individual channels at MINER experimental site, **b** Combined phonon energy distribution with Al fluorescence peak fitted with gaussian, **c** Baseline resolution is measured at MINER experimental site is ~61 eV

as the detector is kept in a Cu casing. During testing in the laboratory (non-reactor environment) with ^{241}Am source, the detector provides a baseline resolution of 39.6 eV [5]. Because of its excellent baseline resolution, we expect to see Al fluorescence at 1.49 keV and Cu fluorescence at 8.05 and 8.9 keV. The Cu fluorescence will be seen mostly in the outer phonon channel (channel A) whereas the Al fluorescence, will be present in all the channels mostly in the bulk of the detector. We do observe the Al fluorescence however we did not detect any Cu fluorescences.

The good events are selected by considering events with good χ^2 values from the pulse template fit. We also consider shared events which deposit energy in the sapphire detector but are consistent with noise in the top and bottom germanium detectors. Those events are known as 'single scatter' events. Another cut is used to select only bulk events using partition variables for outer channel A. The variable is defined as the ratio of the amplitude in channel A to the sum of the amplitude in all the channels: $\frac{A}{A+B+C+D}$. Events with partition value <0.15 are selected. After applying these cuts, a prominent peak is found in individual sapphire channels as well as in the combined spectrum. Figure 4b shows the Al fluorescence peak in the combined spectrum at 1.49 keV and it is fitted with a gaussian (red line) for calibration. Using the calibration, we have calculated the baseline resolution using noise data taken with random triggers. A baseline resolution of ~61 eV is observed at the reactor environment (see Fig. 4c). This is higher compared to a similar measurement with the sapphire detector at the test facility (~39.6 eV). This maybe due to the different environmental conditions causing higher electronic noise. Rare events like dark matter or neutrino hardly interact with the detector material. On the other hand, most of the backgrounds are multiple scatter. For rare event search experiments single scatter background rate in the detector should be ~50 DRU where the DRU stands for Differential Rate unit or counts/kg-keV-days, is used mostly to quantify event rate in rare event search experiments. In MINER, with only Lead and water bricks shielding we measure the single scatter rate in sapphire to be ~1200 DRU in the energy interval 50 to 110 keV in reactor off condition. MINER is recently optimizing the experimental shielding design to reduce the background rate as low as 50 DRU for it's science run aiming for CEνNS measurement.

5 Conclusion and Outlook

Performance of sapphire detector at the MINER experimental site has been shown from the recent data. The detector shows a baseline resolution of ~39.6 eV at test facility which equates a conservative threshold of ~75 eV (<100 eV desired for CEνNS search). Although, at the experimental site it shows baseline resolution of ~61 eV which is higher than at test facility due to reactor environment causing more noise. From the reactor off data Al fluorescence has been identified. The single scatter background rate in the detector is ~1200 DRU in the energy range 50–110 keV. MINER plans to take more engineering runs with different payload using sapphire with Ge detector and Si HV detector in coincidence. Due to having very low baseline resolution, the detector could be a excellent candidate for CEνNS and low mass dark matter searches.

Acknowledgements We acknowledge the seed funding provided by the Mitchell Institute for the early conceptual and prototype development. We would like to acknowledge the support of DAE-India through the project Research in Basic Sciences - Dark Matter and SERB-DST-India through the J. C. Bose Fellowship.

References

1. D.Z. Freedman, Coherent effects of a weak neutral current. Phys. Rev. D **9**, 1389–1392 (1974)
2. A. Drukier, L. Stodolsky, Principles and applications of a neutral-current detector for neutrino physics and astronomy. Phys. Rev. D **30**, 2295–2309 (1984)
3. V. I. Kopeikin, V. V. Sinev, Energy spectrum of reactor antineutrinos and searches for new physics (recent developments). Particles Nucl. Lett. (5)[108] (2001)
4. F. Caleb, W. Sam, Getpy documentation (2021)
5. S. Verma, S. Maludze, M. Lee, M. Chaudhuri et al., Large-mass, low-threshold sapphire detector for rare event searches. NIMA **1046**, 167634 (2023)

Study of Neutron Response Using Time of Flight Technique in ISMRAN Detector

Roni Dey, Pawan Kumar Netrakanti, Dipak Kumar Mishra, Shiba Prasad Behera, Raman Sehgal, Vishwajeet Jha, and Lalit Mohan Pant

1 Introduction

ISMRAN detector setup is designed to measure the yield and energy spectrum of $\overline{\nu}_e$, via the inverse beta decay (IBD) process, for monitoring the reactor thermal power and fuel evolution and also search for the existence of sterile neutrino with a mass on the order of $\sim$1 eV /c^2. The excess of $\overline{\nu}_e$ events in data compared to the predictions particularly at the energy range between 5 and 7 MeV in the measured positron energy spectrum will also be addressed using ISMRAN array. The detector setup at Detector Integration Laboratory (DIL) in BARC, consisted of 90 PSBs, arranged in the form of a matrix in an array of 9×10 in non-reactor environment. Each PSB is wrapped with Gadolinium Oxide (Gd_2O_3) coated on aluminized mylar foils, 100 cm long with a cross-section of 10×10 cm^2 [1]. Three inch diameter, PMTs are coupled at the both ends of each PSB for signal readout of the triggered events. The data acquisition system (DAQ), CAEN VME based 16 channels and 14 bits waveform digitizer (V1730) of high sampling frequency 500 MS/s has been used for the pulse processing and event triggering from each PSB independently. The anode signals from PMTs at both ends of a PSB are required to have a time coincidence of 20 ns to be recorded as a triggered event. The timestamped data from each PSB is then further analyzed offline using energy deposition, timing and position information to build an event. In this paper, we present the measurement of scintillation-light yield of fast neutron due to the recoiling protons in PSB, which has been determined by converting TOF spectrum to neutron kinetic energy to get the proton recoil parametrization of PSB. Fast neutron capture time distribution in ISMRAN array has also been discussed in detail.

R. Dey (✉) · P. K. Netrakanti · D. K. Mishra · S. P. Behera · R. Sehgal · V. Jha · L. M. Pant
Bhabha Atomic Research Centre (BARC), NUCLEAR PHYSICS DIVISION (NPD), Trombay, Mumbai 400085, MH, India
e-mail: rnd48388@gmail.com

R. N. Patra (ed.), *Advanced Radiation Detector and Instrumentation in Nuclear and Particle Physics*, Springer Proceedings in Physics 282,
https://doi.org/10.1007/978-3-031-19268-5_14

2 Results and Discussion

2.1 *Fast Neutron Energy Response in the ISMRAN Array*

Fast neutron energy response in PSB has been measured using americium-beryllium (^{241}Am-^{9}Be) neutron source. Fast neutrons are produced via $^{9}\text{Be}(\alpha, n)^{12}\text{C}$ reaction, where the α particle is produced in the radioactive decays of ^{241}Am. In about 60% of the cases, the carbon nucleus is produced in an excited state, and emits a 4.438 MeV γ-ray in addition to the neutron. We used a 2$''$ cerium bromide ($CeBr_3$) detector for triggering 4.438 MeV γ-ray, which provides a reference start time for the corresponding emitted neutrons. The $CeBr_3$ detector has been placed close to the ^{241}Am-^{9}Be source. The ISMRAN array was located 1.1 m away from the ^{241}Am-^{9}Be source at the source height, as shown in Fig. 1. The $CeBr_3$ detector is calibrated using standard radioactive γ-rays sources and the energy resolution is obtained 3.8% at 0.662 MeV. As a trigger for the start time, 4.438 MeV γ-ray has been tagged in $CeBr_3$ detector and neutron or γ-ray as stopped time candidate is recorded at the first column (10 PSBs) of ISMRAN array. The time coincidence window between start and stop detectors is chosen to be 100 ns. By recording the start and stop time signals from $CeBr_3$ and first column of ISMRAN array, the TOF is reconstructed for discrimination between the γ-rays and neutrons on the first column of the ISMRAN array. Figure 2a shows the comparison of energy deposited untagged γ-rays spectrum from ^{241}Am-^{9}Be source and the γ-rays from natural background (without source) in $CeBr_3$ detector. The peaks in the γ-rays distribution in $CeBr_3$ detector for ^{241}Am-^{9}Be source between 3.3 and 5 MeV correspond to the neutron-associated γ-rays from the de-excitation of $^{12}C^*$. As it can be seen from Fig. 2a, a full-energy peak at 4.438 MeV corresponds to high energy γ-ray due to the de-excitation of carbon nucleus from neutron source while the corresponding first and second escape peaks appear at ~3.95 MeV and ~3.5 MeV, respectively. Figure 2b displays the TOF distribution of fast neutron as a function of γ energy deposition in $CeBr_3$ detector within the

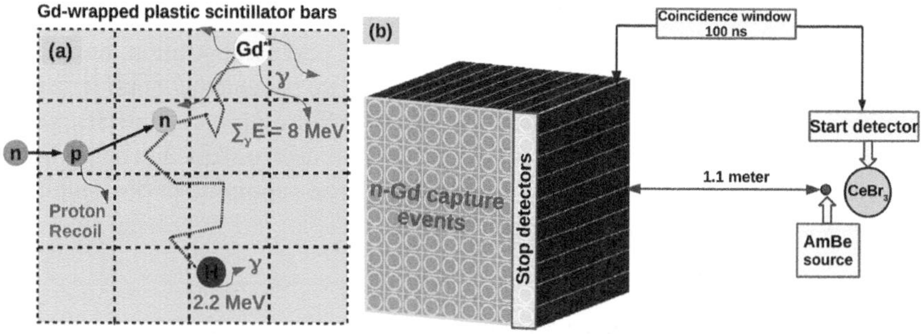

Fig. 1 Panel (a) shows the schematic representation of fast neutron event mimicking prompt and delayed event signatures in ISMRAN array. Panel (b) shows the schematic representation of TOF experimental set up at DIL

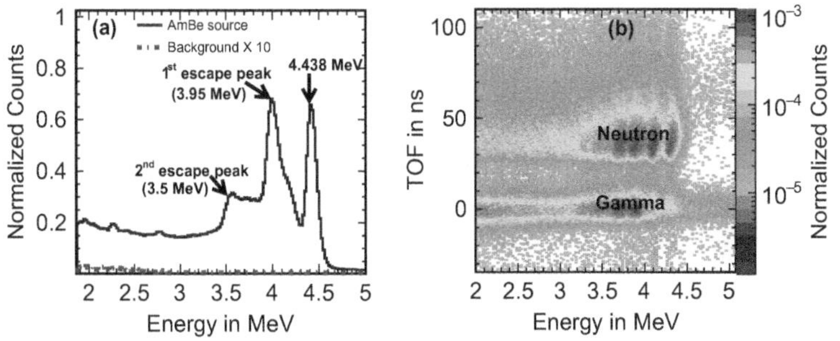

Fig. 2 Panel (a) shows the comparison of energy deposited untagged γ-rays spectrum from ^{241}Am-^{9}Be source and the γ-rays from natural background (without source) in $CeBr_3$ detector. Panel (b) shows the TOF versus γ energy deposited spectrum in $CeBr_3$ detector within the time coincidence of 100 ns between PSBs and $CeBr_3$ detector

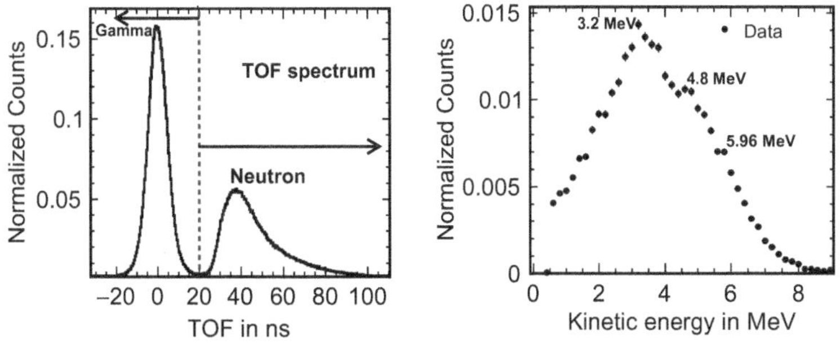

Fig. 3 Panel (a) shows the projected distribution of TOF spectrum for γ-ray and neutron. Panel (b) derived the kinetic energy distribution of fast neutron from ^{241}Am-^{9}Be source using TOF technique

time coincidence window of 100 ns. Two bands at $\sim$ 1 ns and $\sim$ 40 ns in the TOF distribution are due to γ-rays and fast neutron, respectively. The separation between γ-rays and neutrons is excellent and this feature is used to tag the fast neutrons on the first column of the ISMRAN array.

Figure 3a shows the projected distribution of TOF for gamma and neutron by tagging high energy gamma 2.0 to 5.0 MeV in $CeBr_3$ detector. A peak at $\sim$ 1.0 ns correspond to the γ-rays, having width of $\sim$4.5 ns. T_0, the instant of emission of the neutron from the source, was determined from the location of the γ-rays in the TOF spectrum using the speed of light (c) and the measurements of the distance (L) between the PSBs and the source. The γ-rays and fast-neutron distributions are clearly identified. The kinetic energy distribution of the neutrons can be determined by TOF spectrum using following classical expression of neutron kinetic energy. Figure 3b shows the neutron kinetic energy distribution, derived from the TOF spectrum of tagged neutron between 20 and 100 ns.

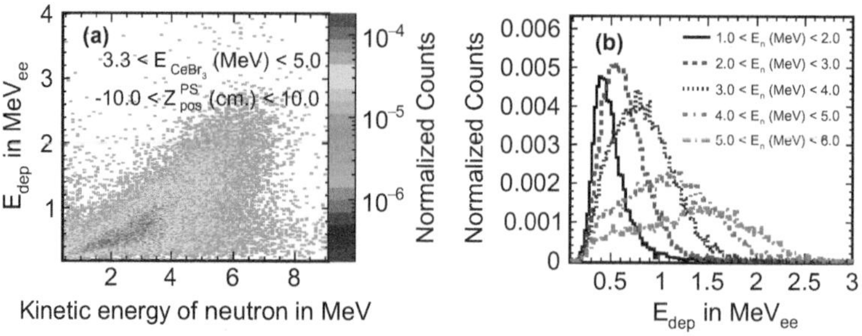

Fig. 4 Panel (a) shows the energy deposition (E_{dep}) of tagged neutrons in PSBs versus kinetic energy of tagged neutrons. Panel (b) shows the projected distribution of neutron energy deposition (E_{dep}) for different kinetic energy bins of tagged neutrons

$$E_n = \frac{1}{2}mV^2 = \frac{1}{2}m\left(\frac{L^2}{t^2}\right) = \alpha^2\left(\frac{L^2}{t^2}\right); \; where, \alpha = 72.3(\sqrt{ev}).\mu s/m, \quad (1)$$

$$TOF_n = \frac{72.3\,L}{\sqrt{E_n}}; \; TOF_\gamma = \left(\frac{L}{c}\right). \quad (2)$$

Figure 4a shows the energy deposition (E_{dep}) by fast neutrons, in MeV_{ee}, in PSB as a function of kinetic energy of neutron derived from TOF. To minimize the contributions from the accidental natural background, γ-ray energy selection between 3.3 and 5.0 MeV is made which covers the first escape, second escape and photoelectric peak for 4.438 MeV γ-ray from ^{241}Am-^{9}Be source. The projection of the E_{dep} of the neutron in PSB for different neutron kinetic energy bins is shown in Fig. 4b. The width of the projected E_{dep} distribution increases with increasing kinetic energy of the neutron. This is due to the fact that exact binning in the TOF distribution for deriving kinetic energy is not possible and the smearing effect on kinetic energy is observed more prominently towards smaller TOF values which yields larger neutron kinetic energies. To reduce this smearing effect, the projection of E_{dep} by the fast neutrons in PSBs are plotted in bins of 1 MeV for the derived kinetic energy of the fast neutrons. Also for fast neutrons with higher energies, the multiple scattering within PSBs can be characterized by broader signal. To get the estimation of scintillation-light yield of neutron for recoiling protons in PSB, parametrization has been done between kinetic energy of neutron and E_{dep} due to recoiling proton in PSB with the following empirical formula, which is represented in Eqs. 3 [2] (Fig. 5),

$$E_{dep} = AE_n - B\left(1 - e^{(-CE_n^D)}\right) \quad (3)$$

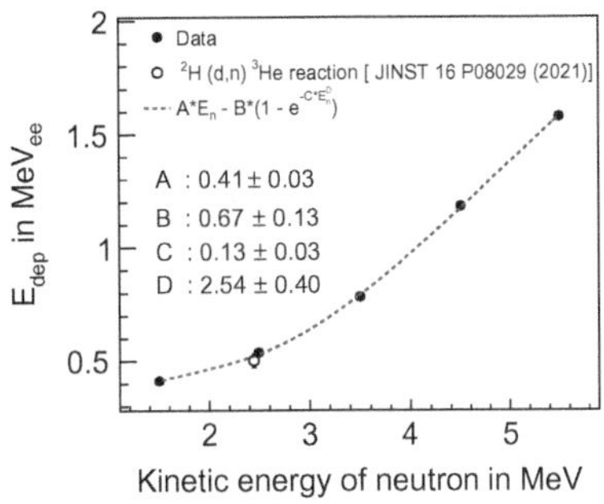

Fig. 5 Parametrization of E_{dep} due to proton recoils as a function of neutron kinetic energy in PSB

2.2 *Measurement of Neutron Capture Time Distribution in the ISMRAN Array*

We have also measured the capture time distribution of fast neutron in ISMRAN detector setup. TOF technique was employed to discern between γ-rays and fast neutron from ^{241}Am-^{9}Be source in the first column (10 PSBs) of ISMRAN array. By tagging the neutron in this column of PSBs as a prompt event we searched the n-Gd capture event as a delayed candidate event in rest of the ISMRAN detectors within a coincidence capture time window of 1000 μs. Only those PSBs are selected for the sum energy (E_{sum}) of delayed candidate events where the individual energy deposition in each PSB is between 0.25 and 10.0 MeV, the E_{sum}, is required to be in the energy range of 2.6 to 10.0 MeV and the number of bars hit (N_{bars}) should be in the range of 4 to 14. All the selection criteria used for the searching of stop events (delayed events) in ISMRAN array are benchmarked with GEANT4 based monte carlo simulation for fast neutron from ^{241}Am-^{9}Be source in ISMRAN array. Figure 6a displays the n-Gd capture time (ΔT_{cap}) distributions of neutrons and γ tagged events. The black solid points represents ΔT_{cap} distribution for all the prompt-delayed pairs reconstructed from neutron tagged events within the time window of 1000 μs. One the other hand, the ΔT_{cap} distribution for γ tagged events (red square) shows a uniform distribution in ΔT_{cap} indicating the randomness in the prompt-delayed event pairs, which scaled with the neutron events above $\Delta T > 300$ μs. Figure 6b shows the ΔT_{cap} distribution of neutron tagged events, which is fitted with a combined function consisting of an exponential term for the neutron thermalization and capture time in PSBs and a constant term representing the accidental residual background. For fast neutron, the fit results in a characteristic time (τ) of 68.29 ± 9.48 μs, which is very similar to the characteristic capture time of thermal neutron for IBD delayed events. This way we have demonstrated a novel technique for the determination of the neutron capture time on Gd in ISMRAN array inspired by the data driven method.

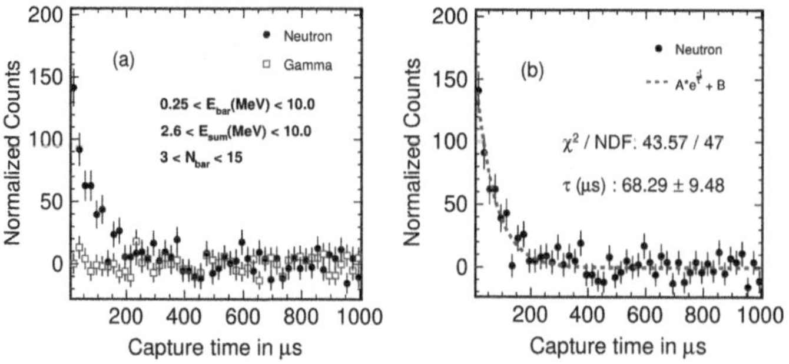

Fig. 6 Panel (a) shows the measured ΔT_{cap} distribution for the start obtained from the first column of the ISMRAN using the tagged fast neutron or γ-ray events with stop recorded in the rest of the ISMRAN array. Panel (b) shows the fit result for ΔT_{cap} distribution between tagged fast neutron from first column in ISMRAN array and stop from rest of the ISMRAN array

3 Summary

The fast neutron energy response in ISMRAN detector is studied with ^{241}Am-^{9}Be source using TOF technique. This technique enabled the mapping of the response of the PSB to the fast neutrons as a function of their kinetic energy, which is useful to get the scintillation light yield (due to recoiling protons) parametrization for PSB. These results also indicate the capture time distribution of fast neutron are indistinguishable from those of $\overline{\nu}_e$ events. For separating the prompt IBD events from fast neutron background can be achieved by using the segmented geometry of ISMRAN array and combining energy dependent variable such as energy ratios with other topological event selection cuts in PSBs along with the implementation of an advanced machine learning algorithms.

References

1. P.K. Netrakanti et al., Measurements using a prototype array of plastic scintillator bars for reactor based electron anti-neutrino detection. Nucl. Instrum. Methods Phys. Res. Sect. A. **1024**, 166126 (2022). https://doi.org/10.1016/j.nima.2021.166126
2. R. Dey et al., Characterization of plastic scintillator bars using fast neutrons from D-D and D-T reactions. J. Instrum. **16**, P08029 (2021). https://iopscience.iop.org/article/10.1088/1748-0221/16/08/P08029

A Compact and Cost Effective Data Acquisition Module (C-DAQ) for Particle Physics Instrumentation

Yuvaraj Elangovan, M. N. Saraf, B. Satyanarayana, Sagar Sonavane, S. Thoi Thoi, R. R. Shinde, and G. Majumder

1 Introduction

FPGA based data acquisition modules are used for particle and nuclear physics experiments [1]. They are used to acquire signals from detectors with high channel density. Also FPGAs ability to process several signals simultaneously makes it more suitable for high speed instrumentation. But FPGAs are slightly larger form factors for small detector setups. Currently most of the prototyping and small detector instrumentation are made using commercially available DAQs [2] which are expensive or NIM modules which requires lots of wires and physical space. In this paper we are proposing a novel idea of a compact and cost effective DAQ solution using FPGA for small detector prototyping and instrumentation.

C-DAQ is a compact and cost effective FPGA based Data Acquisition module Fig. 1. It consists of a low form factor Max10 FPGA from Intel. It can handle eight negative polarity channels suitable for Photomultiplier Tube (PMT) signals. Each of these channel signal paths consist of negative reference discriminators. All the eight channel signals are discriminated above noise level using adjustable potentiometer and driven to the FPGA. C-DAQ consist of two units motherboard and a daughter card Fig. 2.

Y. Elangovan (✉) · M. N. Saraf · B. Satyanarayana · S. Sonavane · S. Thoi Thoi · R. R. Shinde · G. Majumder
Department of High Energy Physics, Tata Institute of Fundamental Research, Homi Bhabha Road, Navy nagar, Colaba, Mumbai 400005, MH, India
e-mail: e.yuvaraj@tifr.res.in

R. N. Patra (ed.), *Advanced Radiation Detector and Instrumentation in Nuclear and Particle Physics*, Springer Proceedings in Physics 282,
https://doi.org/10.1007/978-3-031-19268-5_15

Fig. 1 C-DAQ compact data acquisition module

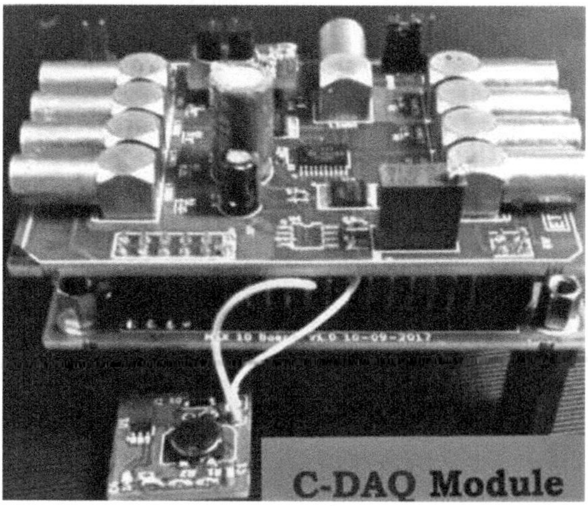

Fig. 2 Mother board and daughter board

2 Daughter Card

Daughter card Fig. 3 was designed to accept eight PMT signals. These signals have pulse heights varying from $-10\,\text{mV}$ to $-1\,\text{V}$. These signals are received by daughter card using LEMO board mount connectors. These input signals are connected to two quad channel MAX9108 [3] comparator ICs. Whenever a PMT signal crosses the reference voltage for the comparator, an output TTL pulse is generated and fed to FPGAs 3.3 V TTL configured input pins. The reference voltage for the comparator is supplied by an on board negative power supply. This power unit converts a positive 3.3 V supply to negative 3.2 V supply. The Discriminated signals are then driven using a driver to the FPGA motherboard via Connecters. Apart from signal discrimination the daughter card also houses a TTL(0 to 3.3 V) to NIM (0 to $-800\,\text{mV}$) converter used to convert FPGA generated trigger signal to NIM format. This can be used for interfacing another DAQ system.

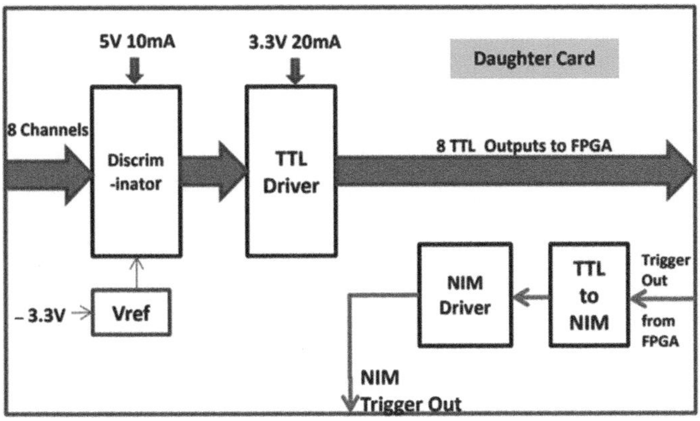

Fig. 3 Daughter card block diagram

3 FPGA Mother Board

Max10 FPGA Evaluation board Fig. 4 developed for workshop and outreach activities is modified as a motherboard. The evaluation board presents a compact and low priced FPGA development platform suitable for newcomers to the FPGA world. The on-board Intel MAX 10 FPGA revolutionizes non-volatile integration by delivering advanced processing capabilities in a low-cost, single chip small form factor programmable logic device. The board is designed to be used in the simplest possible implementation targeting the Intel MAX 10 device up to 2000 LEs. The MAX10 Eval board [4] has a collection of interfaces including two external GPIO headers to extend designs beyond the MAX10Board, on-board USB-to-UART device for interfacing to a PC, as well as general user peripheral with LEDs, 7-segment displays and push-buttons Fig. 5.

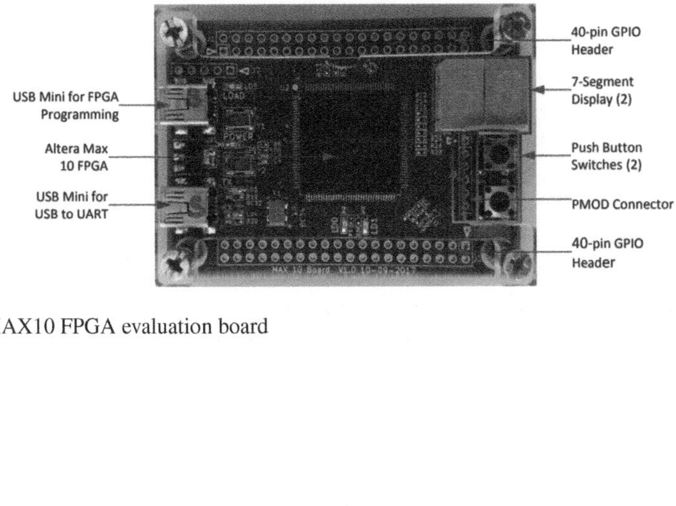

Fig. 4 MAX10 FPGA evaluation board

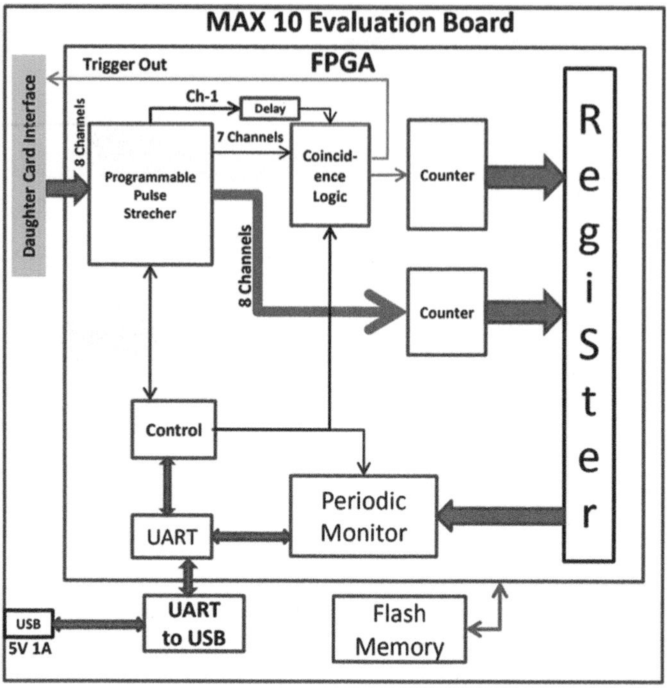

Fig. 5 FPGA logic

4 Data Acquisition

The MAX 10 FPGA configuration for C-DAQ consists of eight 32 bit counters dedicated for each input signal. For better counting all the eight signals are stretched to 100 ns separately using eight moonshots with a system clock of 50 MHz. These counters are incremented when the respective channel input signal arrives. The resolution of the counter is 20 ns, with accuracy of 40ns. Because 20 ns jitter is added due to the monoshot. A programmable coincidence trigger logic is implemented to create a trigger output. When the input signals coincide with each other and if they match with a pre-programmed coincidence pattern, a 200 ns pulse is generated in the trigger output of the FPGA. To count the coincident events a separate 32 bit counter is used. In the daughter card the TTL trigger output from FPGA is converted to NIM signal to facilitate external DAQ interfaces. A dedicated UART logic transfers all the counter values to the server (PC/Laptop) via USB interface. A dedicated USB to serial converter CH340G [5] is used for connecting to the COM port of the server. The data acquisition is triggered by the periodic monitoring logic which generates a 100 ns pulse every second. On every monitoring trigger the eight 32 bit counters values and Coincidence event counter values are transported to the server via UART

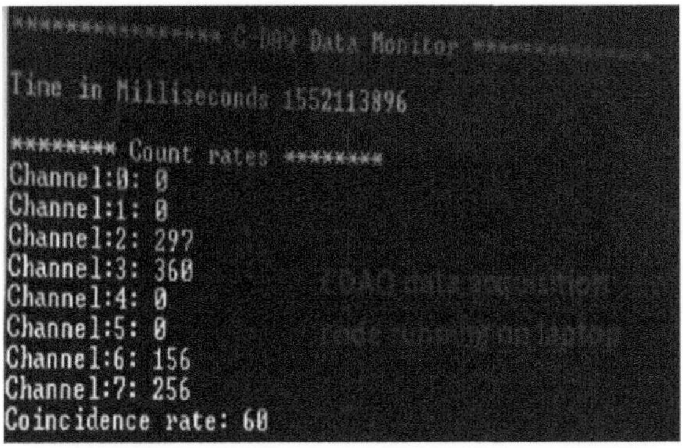

Fig. 6 Data acquisition display

serial logic runs at 115200 baud rate. A simple python script was designed to read these values to store and update on screen Fig. 6 . Also this python script can control and configure the DAQ logic via UART commands.

5 Calibration and Testing

To calibrate the module a fixed frequency tail pulse generator was used. A300 mV negative pulse generated with a frequency 10 Hz is fed to the C-DAQ module. With a monitoring window of 10 s one of the channel count rates was measured Fig. 7. Also counters inside the FPGA are tested with an inbuilt pulse generator logic. A linear response of the DAQ was observed. To check the performance of the module, we prepared a simple Muon telescope of 4 plastic scintillators coupled with Hamamatsu PMTs [6]. The scintillators are all of dimensions 96 × 32 cm. All the 4 scintillator signals are connected to C-DAQ and the module was interfaced with a laptop via

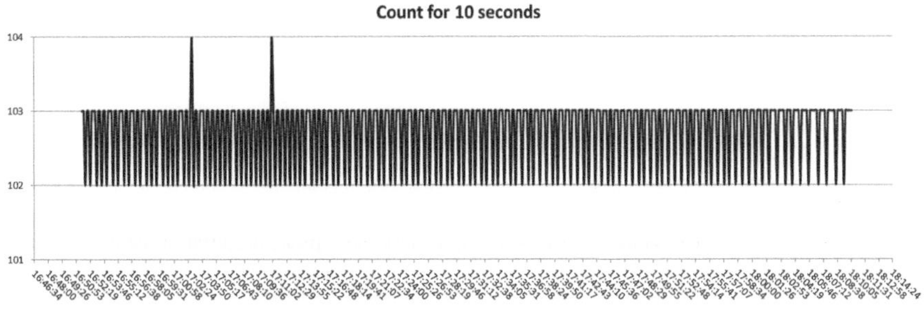

Fig. 7 Counts observed with a tail pulse generator

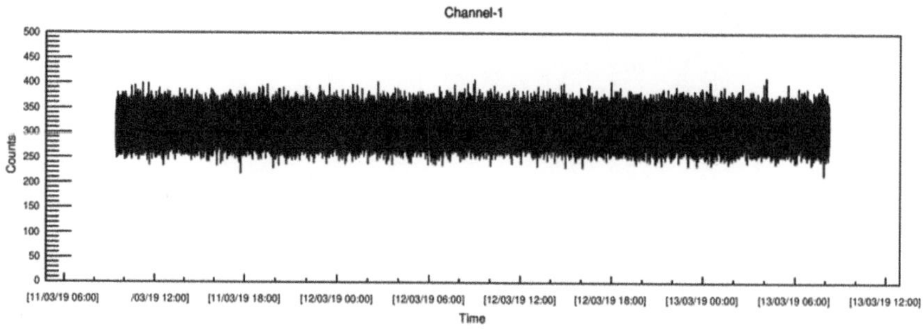

Fig. 8 Noise rates recorded from one of the plastic Scintillator

USB cable. A continuous data acquisition was done for a period of 48 h and the individual channel count values are plotted. One of the channel count rates is shown in Fig. 8. A stable count rate was observed without any DAQ interruptions.

6 Conclusion

Presently only count rates are being collected by the server. Temperature, Pressure, Humidity (TPH) sensor readout need to be added in the existing logic. Adding amplifiers with both polarities at the input stage may be considered. Autonomous mode and normal modes of operation will be implemented. Also an 8 channel carry chain Time to Digital Converter (TDC) [7] may be implemented for timing studies.

References

1. Development of an FPGA-based Readout for Particle Tracking Proceedings of the DAE Symposium on Nuclear Physics, vol. 64 (2019). http://www.sympnp.org/proceedings/64/G74.pdf
2. Commercially available Compact DAQ Solutions. https://www.caen.it/subfamilies/programmable-daq-platforms
3. Max9108 Comparator Data Sheet. https://datasheets.maximintegrated.com/en/ds/MAX9107-MAX9109.pdf
4. MAX10 Evaluation Board user Manual. https://www.tifr.res.in/~ehep2019/pdfs/expt02_4a_Max_10_Dev_Board_User_Manual_v122.pdf
5. CH340G USB to Serial Converter Data Sheet. https://cdn.sparkfun.com/datasheets/Dev/Arduino/Other/CH340DS1.PDF
6. https://www.hamamatsu.com/eu/en/product/optical-sensors/pmt/pmt_tube-alone.html
7. K. Cui, Z. Liu, R. Zhu, X. Li, FPGA-based high-performance time-to-digital converters by utilizing multi-channels looped carry chains. Int. Conf. Field Program. Technol. (ICFPT) **2017**, 223–226 (2017). https://doi.org/10.1109/FPT.2017.8280146

Author Index

R. N. Patra (ed.), *Advanced Radiation Detector and Instrumentation in Nuclear and Particle Physics*, Springer Proceedings in Physics 282,
https://doi.org/10.1007/978-3-031-19268-5

Printed by Printforce, the Netherlands